About the authors

Graham Lawton

After a degree in biochemistry and an MSc in science communication, both from Imperial College, Graham Lawton landed at *New Scientist*, where he has been for almost all of the twenty-first century, first as features editor and now as staff writer. His writing and editing have won a number of awards.

Stephen Hawking

Stephen Hawking was a brilliant theoretical physicist and is generally considered to have been one of the world's greatest thinkers. He held the position of Lucasian Professor of Mathematics at the University of Cambridge for thirty years and was the author of *A Brief History of Time* which was an international bestseller. His other books for the general reader include *A Briefer History of Time*, the essay collection *Black Holes and Baby Universes, The Universe in a Nutshell, The Grand Design* and *Brief Answers to the Big Questions.* He died on 14 M

How to Be Human

The Brain

Question Everything

Nothing

Chance

How Long is Now?

The Universe Next Door

This Book Will Blow Your Mind

Do Polar Bears Get Lonely?

Does Anything Eat Wasps?

Farmer Buckley's Exploding Trousers

How to Fossilise Your Hamster

How to Make a Tornado

Why Are Orangutans Orange?

Why Can't Elephants Jump?

Why Don't Penguins' Feet Freeze?

Will We Ever Speak Dolphin?

THE ORIGIN OF (ALMOST) EVERYTHING

INTRODUCTION BY

PROFESSOR

STEPHEN HAWKING

WORDS BY

GRAHAM LAWTON

NewScientist

First published in Great Britain in 2016 by John Murray (Publishers)
First published in the United States of America in 2016
by Nicholas Brealey Publishing

Imprints of John Murray Press
An Hachette UK company

This paperback edition published in 2019

1

Additional writing by Sally Adee, Gilead Amit, Colin Barras, Alison George,
Joshua Howgego, Justin Mullins, Mick O'Hare and Richard Webb
Fact checking by Chris Simms
Copy editor: Hilary Hammond
Proofreaders: Miren Lopategui and Margaret Gilbey
Indexer: Caroline Wilding

A CIP catalogue record for this title is
available from the British Library

Paperback ISBN 978 1 473 69626 6
UK eBook ISBN 978 1 473 62926 4
US eBook ISBN 978 1 857 88939 0

Typeset in Minion Pro by Palimpsest Book Production Ltd, Falkirk, Stirlingshire

Printed and bound in Great Britain by Clays Ltd, Elcograf S.p.A.

John ... d
recycla ... ts.
The lo ... n

John Mu ... g, US
Ca ...
50 Vict ... e Street,
Lon ...
www.j ... m

CONTENTS

INTRODUCTION

Professor Stephen Hawking

EXISTENCE: WHERE DID WE COME FROM? Why are we here? According to the Boshongo people of central Africa, before us there was only darkness, water and the great god Bumba. One day Bumba, in pain from a stomach ache, vomited up the Sun. The Sun evaporated some of the water, leaving land. Still in discomfort, Bumba vomited up the Moon, the stars and then the leopard, the crocodile, the turtle and, finally, humans.

This creation myth, like many others, wrestles with the kinds of questions that we all still ask today. Fortunately, as will become clear, we now have a tool to provide the answers: science.

When it comes to these mysteries of existence, the first scientific evidence was discovered in the 1920s, when Edwin Hubble began to make observations with a telescope on Mount Wilson in California. To his surprise, Hubble found that nearly all the galaxies were moving away from us. Moreover, the more distant the galaxies, the faster they were moving away. The expansion of the universe was one of the most important discoveries of all time.

This finding transformed the debate about whether the universe had a beginning. If galaxies are moving apart at the present time, they must therefore have been closer together in the past. If their speed had been constant, then they would all have been on top of one another billions of years ago. Was this how the universe began?

At that time many scientists were unhappy with the universe having a beginning because it seemed to imply that physics had broken down. One would have to invoke an outside agency, which for convenience one can call god, to determine how the universe began. They therefore advanced theories in which the universe was expanding at the present time but didn't have a beginning.

Perhaps the best known was proposed in 1948. It was called the steady state theory, and it suggested that the universe had existed for ever and would have looked the same at all times. This last property had the great virtue of being a prediction that could be tested, a critical ingredient of the scientific method. And it was found lacking.

Observational evidence to confirm the idea that the universe had a very dense beginning came in October 1965, with the discovery of a faint background of microwaves throughout space. The only reasonable interpretation is that this 'cosmic microwave background' is radiation left over from an early hot and dense state. As the universe expanded, the radiation cooled until it became just the remnant we see today.

Theory soon backed up this idea. With Roger Penrose of Oxford University, I showed that if Einstein's general theory of relativity is correct, then there would be a singularity, a point of infinite density and space–time curvature, where time has a beginning.

The universe started off in the Big Bang and expanded quickly. This is called 'inflation' and it was extremely rapid: the universe doubled in size many times in a tiny fraction of a second.

Inflation made the universe very large, very smooth and very flat. However, it was not completely smooth: there were tiny variations from place to place. These variations eventually gave rise to galaxies, stars and solar systems.

We owe our existence to these variations. If the early universe had been completely smooth, there would be no stars and so life could not have developed. We are the product of primordial quantum fluctuations.

As will become clear, many huge mysteries remain. Still, we are steadily edging closer to answering the age-old questions: Where did we come from? And are we the only beings in the universe who can ask these questions?

PREFACE

Graham Lawton

I HAVE ALWAYS BEEN FASCINATED BY ORIGINS. As a child, I used to go out to the Yorkshire coast with my mum, dad and sister; we'd dig ammonites, belemnites and devil's toenails out of the cliffs and I'd wonder: Where did they come from? What was the Earth like when they were alive?

It wasn't just the natural world that made me ask where things came from. I remember watching the television – probably black and white in those days, but still a technological wonder – and thinking: Who invented that? I couldn't fathom how somebody could have created a box with a screen that projected pictures from far away. Left to my own devices, I thought, I could never have done that.

When I became a science journalist 20 years ago, I realised the powerful pull that origin stories exert on our imaginations. 'Where did we come from?' is one of the most profound and fundamental questions we ask ourselves. (The others are 'How should we live?' and 'Where are we going?', but those are for another day.) I am convinced it is part of human nature to look at something, or ponder some existential question, and say: How did that come to be?

Every society we know of has stories about the origin of the cosmos and its inhabitants. The oldest creation myth on record is the Enuma Elish, written on 2,700-year-old clay tablets and dating from Bronze Age Babylon. But origin stories surely long pre-date that, dating to at least 40,000 years ago when our ancestors became what is known as behaviourally modern humans. To the best of our knowledge, their minds were identical to ours. That means they possessed the capacity for mental time travel – the ability to project themselves into the past and future, allowing our ancestors to transcend the here and now and even the bookends of their own lifetimes to contemplate the deep past and the distant future. They must, like us, have wondered where it all came from.

Perhaps it goes even further back. Maybe even our earliest ancestors had an origin myth, a million-year-old tale relayed in proto-language around a *Homo erectus* campfire. Yes, even origin stories demand an origin story.

The creators of these ancient stories, of course, did not have much to go on: just their immediate experiences and their imagination. More often than not, they fell back on supernatural explanations. Our own culture's origin myth, the Book of Genesis, is one such story. It actually has two bites at the cherry: first the familiar six-day creation myth, and then a slightly different and somewhat contradictory version. Perhaps this is a tacit admission that we can never know for sure, but are driven to have a go.

But add in the power of the scientific method, and mental time travel becomes a precision instrument. We can use telescopes to peer into the early universe and deploy mathematics to understand its properties. Rewinding the clock like this has taken us a very long way indeed – almost to the beginning of the universe itself, as Stephen Hawking explains in his introduction.

Meanwhile, the historical sciences – geology, evolutionary biology and cosmology – allow us to reconstruct events that happened long before humans existed, way back in what is called 'deep time': the birth of our solar system, the origin of life, the evolution of our own species, and many others. Archaeology and history help us to understand our own past and the origin of things for which humans are directly responsible, from early innovations such as cooking to modern technology like the world wide web.

The Origin of (almost) Everything is a compilation of modern origin stories as revealed by science, that brings together the important, interesting and unexpected.

When I started to compile a list of ideas, some were obvious must-haves, such as the Big Bang, the origin of life and the evolution of humans. The rise of human civilisation was another rich seam. Fifteen thousand years ago our ancestors were nomadic hunter-gatherers; now we live in houses, shop in supermarkets and travel around in machines. How did that happen?

Other ideas were less obvious, and I am grateful to my brilliant colleagues at *New Scientist* and John Murray for suggesting some of the more left-field ones: zero, soil and personal hygiene are among my favourites. In the end we had far too much material to squeeze into a single book. The list of ideas that did not make the cut is a long one that includes the origin of cricket and Viennetta ice cream, to name but two. Maybe one day I will write *The Origin of (almost) Everything Else.*

But enough of this mental time travel. I'm very proud of this book. It has been a journey of discovery for me, and I hope it is for you too. Many of the stories it tells changed and evolved as we were working on the book, as new discoveries came to light. That is the restless beauty of science.

My only regret is that the working subtitle did not make it on to the cover (in case you're wondering, it was *From the Big Bang to Belly-button Fluff*, which I think gives you a flavour of its scope). It formally began life in a brainstorm between *New Scientist* and John Murray, but I'd like to think its true origins are on a Yorkshire beach, inside a small boy's head, inspired by the wonder of nature.

But there I go again, travelling in time to try to work out where something got started. We just can't help ourselves.

London, May 2016

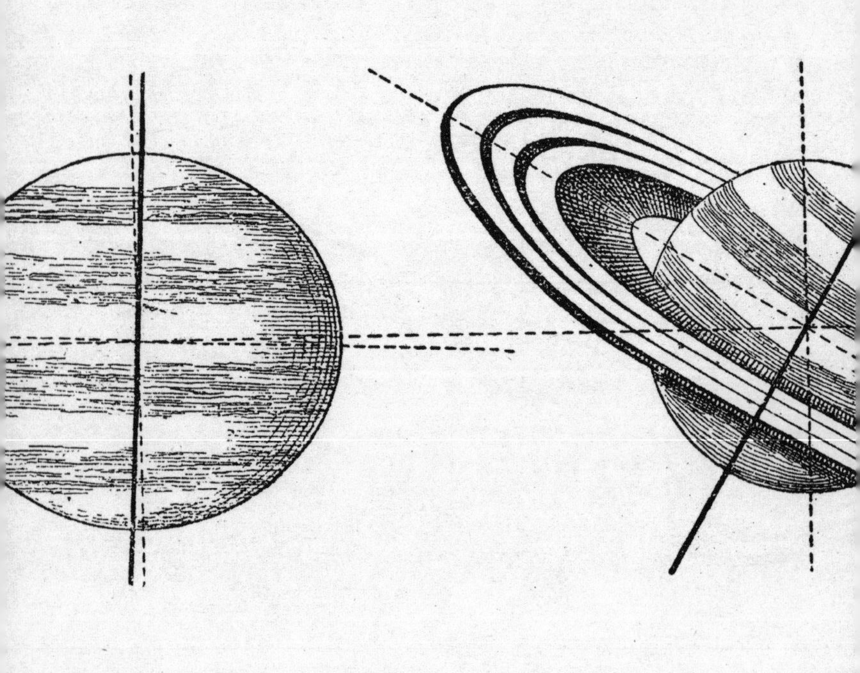

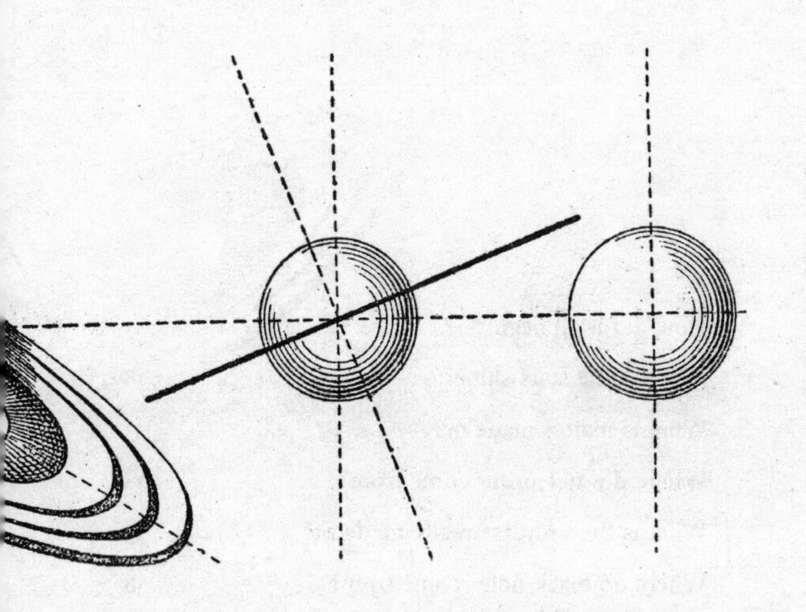

PART ONE

THE UNIVERSE

HOW DID IT ALL BEGIN?

THE UNIVERSE IS BIG. Very big. And yet, if our theory of its origin is correct, then the universe was once small. Very, very small. Indeed, at one point it was non-existent. Around 13.8 billion years ago matter, energy, time and space spontaneously sprang from nothing in the event we know as the Big Bang.

How did that happen? Or to put it another way: What is the origin of everything?

This is the quintessential origin mystery. To most people throughout history, the only plausible answer was 'God did it'. For a long time, even science ducked the issue. In the early twentieth century physicists generally believed that the universe was infinite and eternal. The first hint that it wasn't came in 1929, when Edwin Hubble discovered that galaxies are flying away from one another like shrapnel after an explosion.

The logical conclusion was that the universe must be expanding, and therefore must have been smaller in the past. By imagining the expansion running backwards, like a movie in reverse, astronomers arrived at another logical but rather strange conclusion: the universe must have had a beginning.

The ultimate beginning

At first, many scientists were unhappy with the idea of an ultimate beginning, and so they advanced alternative explanations that did not require one. Perhaps the best known, the steady state universe, was proposed in 1948. According to this hypothesis, the universe had existed for ever and would have looked the same at all times. Astronomers soon found ways to put this claim to the test and found it wanting. Some celestial objects, such as quasars, are only found at great distances away from us, which suggests the universe did *not* look the same at all times. Nonetheless, the steady state theorists did leave a lasting legacy, bequeathing us their expression 'big bang', which was originally coined as a sarcastic dismissal.

The killer blow came in 1965, with the accidental discovery of the faint glimmer of radiation permeating all space. The interpretation of this cosmic microwave background was that it was the 'afterglow' of a universe much hotter and denser than it is today.

These observations were soon backed up by theory. Stephen Hawking and Roger Penrose showed that if general relativity is correct, then there must have been a time when the universe was infinitely small and dense – a moment when time itself began.

The Big Bang is now mainstream science. Cosmologists believe they can trace the evolution of the universe from a split second after its origin to the present day, including a brief period of breakneck expansion called inflation and the birth of the first stars. The actual moment of creation, however, is still a matter of great speculation. At that point our theories of reality start to break down. To make progress, we need to work out how to reconcile general relativity with quantum theory. But, despite decades of hard intellectual graft, physicists are still stuck on that one. Nonetheless, we do have some idea of how to answer the nagging question at the heart of the Big Bang.

How do you get something from nothing?

This is a very reasonable thing to ask, as some basic physics suggests that the universe is overwhelmingly unlikely to exist. The second law of thermodynamics says that disorder, or entropy, always tends to increase over time. Entropy measures the number of ways you can rearrange a system's components without changing its overall appearance. The molecules in a hot gas, for example, can be arranged in many different ways to create the same overall temperature and pressure, so the gas is a high-entropy system. In contrast, you can't rearrange the molecules of a living thing much without turning it into a non-living thing, so that makes us low-entropy systems.

By the same logic, nothingness is the highest entropy state around; you can shuffle it around all you want, and it still looks like nothing.

Given this law, it is hard to see how nothing could ever be turned into something, let alone a universe. But entropy is only part of the story. The other part is a quality that physicists call symmetry – which is not exactly the same as the everyday symmetry we associate with shapes. To physicists, a thing is symmetrical if there is something you can do to it so that after you have finished doing it, it looks the same as it did before. By this definition, nothingness is totally symmetrical: you can do what you like to it and it is still nothing.

As physicists have learned, symmetries are made to be broken, and when they break, they exert a profound influence on the universe.

Indeed, quantum theory tells us that there is no such thing as emptiness. Its perfect symmetry is too perfect to last, broken by a roiling broth of particles that pop in and out of existence.

That leads to the counterintuitive conclusion that, despite entropy, something is a more natural state than nothing. In that sense everything in our universe is just excitations of the quantum vacuum.

Might something similar account for the origin of the universe itself? Quite plausibly. Perhaps the Big Bang was just nothingness doing what comes naturally: a quantum fluctuation that caused an entire universe to pop into existence.

Outside space and time

This, of course, raises the question of what came before the Big Bang, and how long this state of affairs persisted. Unfortunately, at this point common-sense concepts such as *before* become meaningless.

It also raises an even tougher question. This understanding of creation relies on the validity of the laws of physics. But that implies that the laws somehow existed before the universe did.

How can physical laws exist outside of space and time and

without a cause of their own? Or, to put it another way, why is there something rather than nothing?

NOT THE BIG BANG

The Big Bang is the mainstream explanation for the origin of the universe, but it doesn't have the floor to itself. One alternative is that instead of a bang there was a bounce. In this scenario, rewinding our universe takes us through its unimaginably hot, dense beginning and out the other side into the unimaginably hot, dense ending of a previous universe. Another is that the bang was one of many. According to multiverse theory, our universe is just one bubble in a seething froth of universes. Both ideas, however, suggest that the universe had no beginning. That is an even harder concept to grasp than it just popping into existence.

WHY DO THE STARS SHINE?

LOOK INTO THE NIGHT SKY, AND YOU ARE LOOKING BACK IN TIME. The light from Sirius A, the brightest star, takes around eight and a half years to travel across interstellar space to Earth. The most distant star visible to the naked eye, Deneb, is about 2,600 light years away. For all we know, neither star even exists any more.

Look further, and we see even deeper back in time. In 2012 the Hubble Space Telescope published an image called eXtreme Deep Field, created by collecting the faint light from a tiny patch of sky for 23 days. It was festooned with distant galaxies, some so far away that their light was emitted when the universe was only half a billion years old.

The image confirmed what astronomers had long suspected: the universe is essentially the same in all directions, dominated by stars and galaxies not dissimilar to our own. But if Hubble could peer even deeper into the past, it would see a very different universe.

It is now generally accepted that the universe started life as an unimaginably small, dense and hot fireball of matter and energy. This universe did not contain stars and galaxies, and would not do so for another 500 million years.

The oldest galaxy we know of is EGSY8p7, which was born about 600 million years after the Big Bang. Half a billion years later the universe was filled with galaxies, each containing hundreds of billions of stars. How did it go from one extreme to the other?

To answer this question, we have to go back a very long way, to a time just 3×10^{-44}seconds after the Big Bang. This was the start of inflation, a fraction of a millisecond during which the universe expanded exponentially.

Blown up like a balloon

Inflation converted the universe from a seething, roiling knot of matter and energy into something much smoother and more homogenous, a bit like blowing up a crinkled balloon. However, it did not lead to complete uniformity: there were tiny variations from place to place, the stretched-out remnants of the quantum fluctuations that had caused the Big Bang. After inflation ended, the universe continued to expand at a much slower pace, stretching out the variations even further. These were the seeds from which stars and galaxies grew.

We know about them from observations of the cosmic background radiation, a faint glimmer of microwaves permeating all of space which is often called the 'afterglow' of the Big Bang. At first the cosmic microwave background appeared to be the same temperature everywhere: a frigid 2.7 °C above absolute zero. But in 1992 NASA's Cosmic Background Explorer (COBE) satellite mapped it in detail and detected regions where it was slightly colder than average and others where it was slightly warmer.

These differences are tiny – only a few parts in 100,000 – but were enough.

The cold spots correspond to regions of the early universe that contained more matter – largely hydrogen and helium – and were therefore slightly denser than average. Gravitational attraction did the rest, gradually clumping this matter into larger and denser blobs that eventually became so large and dense that nuclear fusion ignited in their cores. The stars were born.

Gravity also accounts for the formation of the clusters of stars we call galaxies and the clusters of galaxies we call, er … galaxy clusters. The latter can be more than 100 million light years across.

Our own galaxy formed this way, and the process continues. The Milky Way, for example, is accreting matter from two nearby satellite galaxies, the Large and Small Magellanic Clouds, and also sucking in gases from space. Already a giant galaxy, far larger and brighter than most others, the Milky Way will eventually become even mightier by merging with another nearby galaxy, Andromeda.

Star formation continues too, in dense regions of interstellar dust known as stellar nurseries. The Hubble Space Telescope has captured dramatic images of huge pillars of gas and dust where newborn stars are emerging from the clouds, complete with protoplanetary discs that will eventually give rise to solar systems. In total the Milky Way spawns about 10 stars a year.

Despite all being born the same way, stars are very varied. Some are bright, others faint; some are blue, others white, yellow, orange or red; some are enormous, others tiny.

Live fast die young

The differences are down to random variations in mass. About 90 per cent of stars are main sequence stars, and they are all doing the same thing: crushing hydrogen nuclei together in their cores to form helium nuclei, a process called fusion. The greater a star's mass, the hotter its centre is and the faster its hydrogen will fuse – so the brighter the star. And the brighter the star, the bluer it is.

A star's mass also dictates how long it will live. Although more massive stars have more fuel to burn, they burn it much faster and die sooner. The most massive stars use up the hydrogen in just a few million years. In contrast, the Sun has been burning for 4.6 billion years and will continue to do so for billions more.

Every main sequence star will someday use up the hydrogen at its centre. It will then begin to burn hydrogen outside its centre while expanding and cooling. It is now a giant or supergiant.

These huge stars lead brief but dramatic lives. They start fusing helium, carbon, neon, oxygen, silicon and sulphur, the last two of which fuse into iron. But iron does not fuse into heavier elements, and at that point the star is doomed to explode as a supernova. Afterwards, the remains collapse into a small but dense sphere. This may be a black hole or a neutron star.

Smaller giants do not explode but just dwindle slowly into hot, dense ghosts called white dwarfs. If enough time elapses, white dwarfs will fade completely and become black dwarfs. But not yet, because the universe is not old enough.

BUILT BY A BLACK HOLE

Galaxies are generally thought to coalesce gradually under the influence of gravity, but there is an alternative and much more dramatic possibility. They may be zapped into existence by high-energy jets of matter slamming into clouds of gas. The jets are unleashed from quasars, extremely luminous objects thought to be powered by supermassive black holes. If correct, this means that the supermassive black holes found at the hearts of most galaxies are the architects of their surroundings rather than the products of them.

WHAT IS MATTER MADE OF?

IMAGINE THAT ON YOUR FIRST BIRTHDAY YOU RECEIVE A RATHER ODD GIFT: A VIAL OF HYDROGEN GAS. The next year you receive some helium, and on your third, a piece of lithium. On your 21st birthday you become the proud owner of some scandium. For your 40th, it's a piece of crystalline zirconium. If you make it to 92, you'll get uranium. But to complete your collection, you'll have to live a lot longer than that.

A hundred and eighteen years, to be precise. That is how many chemical elements we know of: a smorgasbord of solids, liquids, gases, metals and non-metals, some rare, some common, some useful, some not. They are the building blocks of chemistry and life. Where do they all come from?

The facile answer is the Big Bang. But that is not a satisfying one, as the bang itself only produced the three lightest elements: hydrogen, helium and a trace of lithium. What about the rest?

The complete answer requires knowledge of the building blocks of atoms and some basic arithmetic. The simplest atom is hydrogen, made of one proton and one electron. The next simplest are deuterium and tritium, which are hydrogen plus a neutron or two. After that comes helium, with two of everything. Next is lithium, with three. Common sense suggests that larger elements can be made by fusing smaller ones together. And that is exactly how they are formed.

The big squeeze

But it is not quite that simple. Such reactions are difficult to achieve because the two nuclei need a vast amount of energy to fuse. That requires astronomical temperatures: a minimum of 10 million degrees Celsius. There are only two places in the universe that fit the bill: shortly after the Big Bang, and inside stars.

The first phase of element building happened very soon after the Big Bang in an event called nucleosynthesis. Within a hundredth of a second, protons, neutrons and electrons condensed out of the fireball. A few seconds later protons and neutrons began to join forces, driven together by the immense energy of the fireball and glued in place by the nuclear force. These fusion reactions initially formed nuclei of deuterium, which reacted with more protons to produce the stable nucleus of helium.

But that was that. By the time helium appeared the temperature had fallen too low for further fusion to happen to an appreciable degree. A little lithium was probably made, but nothing heavier. Nucleosynthesis was over almost as soon as it began.

About 377,000 years later, business resumed. The temperature dropped to about 3,000 degrees – cool enough for atoms to exist. Hydrogen and helium nuclei mopped up free electrons to form the first complete atoms, elements 1 and 2. While these still make up more than 99 per cent of the visible universe, they are not its only components. To make the heavier, more interesting elements requires stars.

A star is formed when a large mass of gas contracts under its own gravity. Compression raises the temperature in the centre to the point at which nuclei can start to fuse. The first reaction, at about 10 million degrees Celsius, is the fusion of hydrogen nuclei to form helium until the hydrogen is exhausted.

Keep on fusing

What happens next depends on the star's mass. If it is quite small, fusion stops and the core simply becomes a white dwarf. But if it is more massive than eight Suns, fusion continues. Helium nuclei combine to form beryllium (element 4), which reacts with more helium to form carbon and oxygen. In the most massive stars, the core gets so hot that carbon and oxygen fuse further, forming elements as heavy as iron (element 26). The reactions then stop, because iron has the most stable nucleus of all the elements and cannot fuse under these conditions. But in the star's outer layers other nuclear reactions involving neutron capture are gradually constructing ever bigger nuclei, up to bismuth (element 83).

As iron builds up in the core, the star is on borrowed time. It can no longer produce energy by fusion, but gravity is remorseless: it continues to compress the core, raising the temperature to billions of degrees. The centre of the star collapses suddenly; the outer layers fall in, then rebound, spewing the contents of the star out into space in a supernova. The explosion produces a flood of neutrons that creates still more heavy elements, right up to uranium (element 92), the heaviest naturally occurring element found on Earth, and beyond. The supernova expels debris out into space, which eventually becomes incorporated into later generations of stars and planets, including our own.

One exception to a starry origin is the trio of lithium, beryllium and boron. Their nuclei are unstable and are immediately consumed by nuclear reactions in stars. They are rare, but (except for the Big Bang lithium) what little of them exists is believed to have been made from cosmic rays – largeish nuclei travelling through space at high speed. Their energy is so large that when they collide with other atoms, the nuclei can break into smaller fragments.

Artificial elements aside, all the atoms on Earth are either leftovers from the Big Bang or fragments of long-dead stars or cosmic rays. And eventually, when our own star dies, they could be cast back out into space and eventually recondense in a new solar system. How's that for a spectacular comeback?

VERY HEAVY METALS

Elements heavier than uranium were unknown on Earth until the early 1940s, when chemists created plutonium and neptunium by bombarding uranium with neutrons. Since then, 24 more *transuranium elements* have been synthesised in laboratories. The largest yet is oganesson, element 118.

Transuranium elements are often thought of as being entirely artificial, but that is not the case. They are created in supernova explosions, just like ordinary heavy elements. However, they are unstable and tend to fall apart quickly. Naturally occurring ones have completely decayed away since the solar system formed, which is why they do not occur outside laboratories on Earth.

WHERE DO METEORITES COME FROM?

On 15 February 2013 something big exploded high in the sky over Chelyabinsk, just to the east of the Ural Mountains in southern Russia. Most of the object burned up in the atmosphere, but some pieces made it down to Earth. One smashed through the ice of the frozen Lake Chebarkul, leaving a hole 7 metres wide. This was recovered by a diver in October 2013, and weighed in at 570 kilograms. Other much smaller fragments were gathered from all over the region.

Astronomers concluded that the explosion was an asteroid 17 to 20 metres across with a mass of 10,000 tonnes. The initial blast, at an altitude of about 30 kilometres, carried an energy equivalent to 500 kilotonnes of TNT – about 30 Hiroshima bombs. It was the largest extraterrestrial impact on Earth in living memory.

The Chelyabinsk meteorite is now one among more than 30,000 that have been discovered on the surface of the Earth, sometimes immediately after a fall but mostly just lying about on the surface long after the event. Each has an interesting story to tell.

Rocky leftovers

Most meteorites are bits of asteroid, which are themselves leftovers from the formation of the solar system. Asteroids usually sit doing not very much in a belt of rubble between the inner planets and the outer gas and ice giants. But for some reason or other they sometimes get pulled out of orbit or are smashed up and, by chance, end up on a collision course with Earth. These travelling space rocks are called meteoroids.

As soon as they land or are found, meteorites become prized assets for planetary scientists keen to unlock the secrets they hold about the history of the solar system.

The first task is to work out what kind of meteorite it is, which reveals where it probably came from. Meteorite taxonomy is complicated, but broadly speaking there are three groups: stony, iron and stony-iron.

The Chelyabinsk meteorite turned out to be a stony one of a rather common-or-garden type called a chondrite, so called because they contain chondrules – small, round particles of silicate material.

No one knows the origin of chondrules, but they probably started out as globs of molten rock in the cloud of dust and gas that gave birth to the solar system. Around 86 per cent of meteorites are chondrites. They are composed mostly of rock and come from the asteroid belt, meaning they are fairly pristine remains of the material that formed the solar system.

Planet organic

A more unusual class of stony meteorites are the carbonaceous chondrites, so called because they contain unusually high levels of organic chemicals such as amino acids. These meteorites are also thought to be pristine chunks of the primordial material that gave rise to the solar system.

A third class of stony meteorite are the achondrites, so called because they lack chondrules. Around 8 per cent of meteorites fall into this class. Rather than being clumps of primordial material, achondrites appear to be the product of the early stages of planet building, when material accreted together under the influence of gravity to form protoplanets. As they grew bigger and hotter, the protoplanets started to melt. This destroyed the chondrules and also caused heavy elements such as iron and nickel to sink towards the centre, leaving behind a rocky mantle. This outer layer appears to be the source of most achondrites; they are the remains of failed planets that never made it big.

A small handful of achondrites have an even more distinguished origin: they were once parts of the Moon or Mars.

About 1 in 20 meteorites belong to the iron group. They are made largely of iron and nickel and are also the remnants of planet building – fragments of the metal-rich cores of protoplanets that were later smashed to smithereens by collisions. These chunks of space metal help us to understand how our own planet separated into core, mantle and crust.

Stony-iron meteorites, the final broad group, are a slightly unsatisfying halfway house between stony and iron. These rare rocks – just 1 per cent belong in this category – also seem to be derived from the interior of failed planets, close to the boundary between the iron core and the rocky outer layers.

Finding a meteorite is not easy. They are easiest to spot in barren places: Antarctica is especially productive, as the landscape is white and the churning of glaciers concentrates them at the bottom of mountains.

Watch your head

If you do find one, chances are it comes from a large asteroid that broke up about 470 million years ago. This gave rise to a hail of chondrites that rained down on Earth during the Ordovician period. Most of the fragments are still out there and, even now, they make up the majority of meteorites which fall to Earth.

Meteorites occasionally hit people, but there have been no confirmed deaths. In November 1954 a meteorite crashed through the roof of a house in Alabama, bounced off a piece of furniture and hit 34-year-old Ann Elizabeth Hodges on the side. She was badly bruised but made a full recovery. In August 1992 a shower of meteorites fell on Mbale, Uganda. One hit a tree and rebounded on to the head of a boy, but he was unhurt.

PIECES OF THE MOON AND MARS

Between 1969 and 1976 US and Soviet space missions brought about 380 kg of lunar rock back to Earth. But these are not the only Moon rocks on Earth. Vast quantities have also arrived in the form of meteorites, presumably blasted off the surface by impacts.

Mars, too, regularly throws rocks at Earth. Around 130 meteorites are from Mars, the only pieces of the Red Planet that we can hold in our hands. The most famous is ALH 84001, found in Antarctica. In 1996 NASA scientists made the sensational claim that it contained the fossilised remains of Martian bacteria. Sadly, the scientific consensus is that the evidence doesn't stack up to a conclusive case for aliens.

WHAT IS THE UNIVERSE REALLY MADE OF?

THERE'S MORE TO THE UNIVERSE THAN MEETS THE EYE. A lot more. In fact, as far as most of the universe is concerned, you're weird and inconsequential. The everyday stuff that constitutes you and everything you care about makes up less than 10 per cent of it; the rest is built of mysterious entities called dark matter and dark energy. Together, they constitute one of the greatest cosmological mysteries of our time. What they actually are, though, is anyone's guess.

The first of these inconvenient entities to thrust itself on to the scene was dark matter. Back in the early 1930s, Dutch astronomer Jan Oort spotted some anomalies in the way stars orbited within the Milky Way. The only way to explain their behaviour was to imagine that some dark, invisible matter filled the greater part of space.

Swiss astronomer Fritz Zwicky later observed some similar anomalous behaviour in a galaxy cluster 320 million light years away. He found that the galaxies were orbiting each other far faster than gravity said they should, based on the combined masses of their stars. Either the galaxies had to contain much more matter than was visible, or else Newton's law of gravity was wrong. Zwicky opted for the former, and put it down to vast swathes of unseen gas.

In a spin

In the 1970s astronomers made a similar observation of individual galaxies, which turned out to be spinning so fast that they should be ripping themselves apart. They initially plumped for Zwicky's unseen gas explanation, but they ran into trouble. If the unseen stuff was normal matter made of protons, neutrons and electrons, then our understanding of how stars and galaxies form must be wrong: they would never have collapsed quickly enough to form the first stars and galaxies.

So they began to think that something else was out there, a mysterious form of matter that does not absorb or emit light or other electromagnetic radiation, which is why we cannot see it. But it interacts with gravity, which is why we can see its effects on ordinary matter. They called it dark matter.

Cosmologists now believe that dark matter is a significant ingredient of the universe, making up about 27 per cent of it. Without the extra gravity it supplies, galaxies would not form quickly enough, nor form the clusters and superclusters we observe today.

Dark matter is largely concentrated in spherical haloes around galaxies. In fact, most of the mass of a spiral galaxy like our own Milky Way is not contained in stars and planets but in the invisible stuff surrounding them.

Bring on the WIMPs

The frustrating truth, however, is that we still don't know what it is. According to our best theories, it is made of hypothetical particles called WIMPs (weakly interacting massive particles). If correct, then trillions of them must pass through the planet every second. Numerous experiments have tried to detect WIMPs or produce them in a lab, but none has succeeded.

And the more detailed the astronomical observations become, the darker things get. Sometimes there seems to be too much dark matter, as in the case of the dwarf galaxies orbiting the Milky Way. These rotate so quickly they must be chock-full of it. But this is exactly the opposite of what we understand from our theory of galaxy formation, which says we should expect the amount of dark matter in galaxies to be roughly proportional to their size.

At other times, we see too little dark matter. Across the universe, there are between a tenth and a hundredth of the number of small galaxies than our theory of galaxy formation predicts. And then there are galaxies that seem to contain no dark matter at all, even though star clusters circling them do seem to be experiencing an extra gravitational pull.

A matter of gravity

The bottom line is that we badly need to know what constitutes dark matter. If it doesn't exist, then our understanding of gravity is wrong. This is unthinkable to most astronomers, who continue to pin their hopes on dark matter and use observations of the way galaxies move and rotate to help pin down its properties.

If ignorance of about 27 per cent of the universe sounds bad, how about knowing absolutely nothing about a further 70 per cent? This was the unenviable position in which cosmologists found themselves in 1998, with the discovery of a bizarre sort of anti-gravity now known as dark energy.

It started with a routine experiment to measure the expansion rate of the universe, which was expected to be slowing down as gravity gradually put the brakes on the Big Bang. The astronomers were searching for supernovae, exploding stars whose light would confirm these details.

The supernovae had a different tale to tell. Distant ones turned out to be much further away than would be expected if the expansion of the universe had been slowing all along. The astronomers were stunned by the inevitable conclusion: instead of slowing down, the universe's expansion was speeding up. But why?

It has become the most troubling question in astrophysics, and one we are no closer to answering. Most physicists think the solution lies with an elusive force, dark energy, which lurks in the emptiness of space and accounts for about 70 per cent of the matter and energy in the cosmos, causing space to expand at an ever-increasing rate.

What exactly is this dark energy? Er … we don't know. But there's no lack of ideas. It might be an energy inherent in the fabric of space itself. It might be an exotic field called quintessence that expands space at changing rates. It might be a modified form of gravity that repels rather than attracts under certain circumstances. It may even be an illusion.

EINSTEIN'S BRILLIANT BLUNDER

The modern concept of dark energy is less than 20 years old, but Albert Einstein invented something very similar in 1917 as an add-on to his general theory of relativity. He realised that gravity would cause the universe to collapse in on itself, so he added a fudge-factor – the cosmological constant, a mysterious anti-gravity force inherent to empty space. He later changed his mind, calling the constant his 'biggest blunder'. We now know he was ahead of his time.

WHERE DO BLACK HOLES COME FROM?

On a clear dark night, go outside and look for the constellation Sagittarius. Lurking somewhere beyond it is a celestial monster that you can be relieved is very far away: a supermassive black hole. You won't be able to see it; it is obscured by dust, not to mention utterly black and about 27,000 light years away. But we're confident it is there, sitting at the centre of our galaxy.

How can we be so sure? And how did it get there?

First things first. Nobody has ever seen a black hole. So how do we know about them?

Black holes are often seen as a twentieth-century discovery, but the idea can be traced back to 1783, when John Michell, a Yorkshire clergyman and amateur philosopher, submitted a speculative paper to the Royal Society in London.

Michell was tackling the problem of how to measure the distance and magnitude of stars (which even now gives astronomers headaches). His starting point was Isaac Newton's corpuscular theory of light, which proposed that light was made of infinitesimally small particles. Michell reasoned that light emitted by a star would be slowed down by the star's gravity. The magnitude of this slowdown could be used to measure the mass of the star, and hence its distance from Earth.

Michell's lengthy paper – which was published in the Royal Society's house journal *Philosophical Transactions* in 1784 – is mainly concerned with how this might be measured from Earth using prisms. But it also contains a flight of fancy.

Michell reasoned that if the star was massive enough, its gravity would be so strong that even light would be unable to escape its clutches. He calculated that the star would need to have a diameter about five hundred times bigger than the Sun to trap light this way. If such an object existed, he wrote, 'its light could never arrive at us'.

This wholly original idea was tangential to Michell's goal, and so he parked it. 'I shall not prosecute it any further,' he wrote.

Michell was true to his word. He died in 1793, apparently without mentioning the idea ever again.

A few years later, French scholar Pierre-Simon Laplace came up with the same idea while speculating about the properties of very large stars. Their gravitational attraction would be 'so great that no light could escape from their surface. The largest bodies in the universe may thus be invisible.'

Laplace may have intended to take the idea further. But in 1804 it was rendered obsolete by a new theory which characterised light not as a stream of particles but as a wave. If so, light would be unaffected by gravity. The idea was forgotten.

SPAGHETTIFICATION

Anything that reaches a black hole's event horizon gets sucked in, never to be seen again. This process is called *spaghettification* – gravity is so strong that it stretches the unfortunate object (a spaceship or astronaut, for example) into a long, thin thread before it is devoured like somebody sucking up a noodle.

Back in black

Things changed again in 1915 with Albert Einstein's general theory of relativity, which redefined gravity as distortions in space–time caused by massive objects such as stars.

The theory made a strange prediction, though Einstein himself didn't spot it. It was pointed out to him by astronomer Karl Schwarzschild, who discovered it during his spare time on the eastern front (he had volunteered for the German army in 1914, aged 40).

Schwarzschild showed that if enough mass was concentrated into a small enough space, the curvature of space–time would become infinite. The result was a 'singularity' – a point in space–time where gravity was so strong that even light could not escape. Einstein was impressed but did not believe such an object could actually exist. Schwarzschild died in 1916 of a disease he contracted in the trenches and his singularity was eventually dismissed as a purely theoretical entity. In 1939 Einstein published a paper that supposedly 'proved' this point, and the matter was laid to rest. For a while, at least.

In the 1950s astronomers started using radio waves to probe deep space and discovered very, very distant objects such as quasars that were so energetic they could only be understood using general relativity. Out of this grew a new appreciation of the physics of very massive objects, and physicists came to accept that singularities could and probably did exist. One key breakthrough was the idea of the 'event horizon'. This is the 'surface' of the black hole, a boundary in space–time where gravity becomes so strong that nothing can escape.

By the end of the 1960s most physicists accepted that black holes were an inevitable consequence of Einstein's theory.

How does a black hole form?

Any star of roughly double or more the mass of our Sun is destined to become a black hole. Such stars have an immense gravitational field which creates an inward pressure. During their lifetimes, stars counteract this through nuclear fusion reactions in their cores. But when they run out of fuel they can no longer resist and begin to scrunch in on themselves in a process called gravitational collapse.

This sometimes leads directly to the formation of a black hole, or else a huge explosion called a supernova, which blows away the outer layers leaving a core. If this is massive enough it will continue to collapse. As the collapsing material gets denser and denser, its gravitational field becomes so strong that it passes the point of no return where not even light can escape. A black hole is born.

The process, however, does not explain the origin of supermassive black holes, which weigh at least 100,000 times as much as the Sun. These may form simply by vast amounts of matter spiralling into an ordinary black hole over stupendous amounts of time. Or perhaps they form by the merger of lots of ordinary black holes. Or perhaps via the collapse of absolutely huge stars in the early universe.

Intriguingly, however, although we have come to generally accept the existence of black holes, nobody has ever seen one. The closest we have come is the recent detection of gravitational waves caused by the collision of two black holes.

Plans are in progress to image a black hole directly, at which point their existence, like the light trapped inside them, will be inescapable.

THE HOLE STORY

No one is entirely sure where the name 'black hole' comes from. It is often attributed to physicist John Archibald Wheeler, who dropped it into a lecture in 1967. But according to the *Yale Book of Quotations*, it first appeared in print in 1964 in a report of a meeting of the American Association for the Advancement of Science. It seems likely that the name was circulating among astrophysicists before Wheeler – who had an ear for a catchy phrase – picked it up and popularised it.

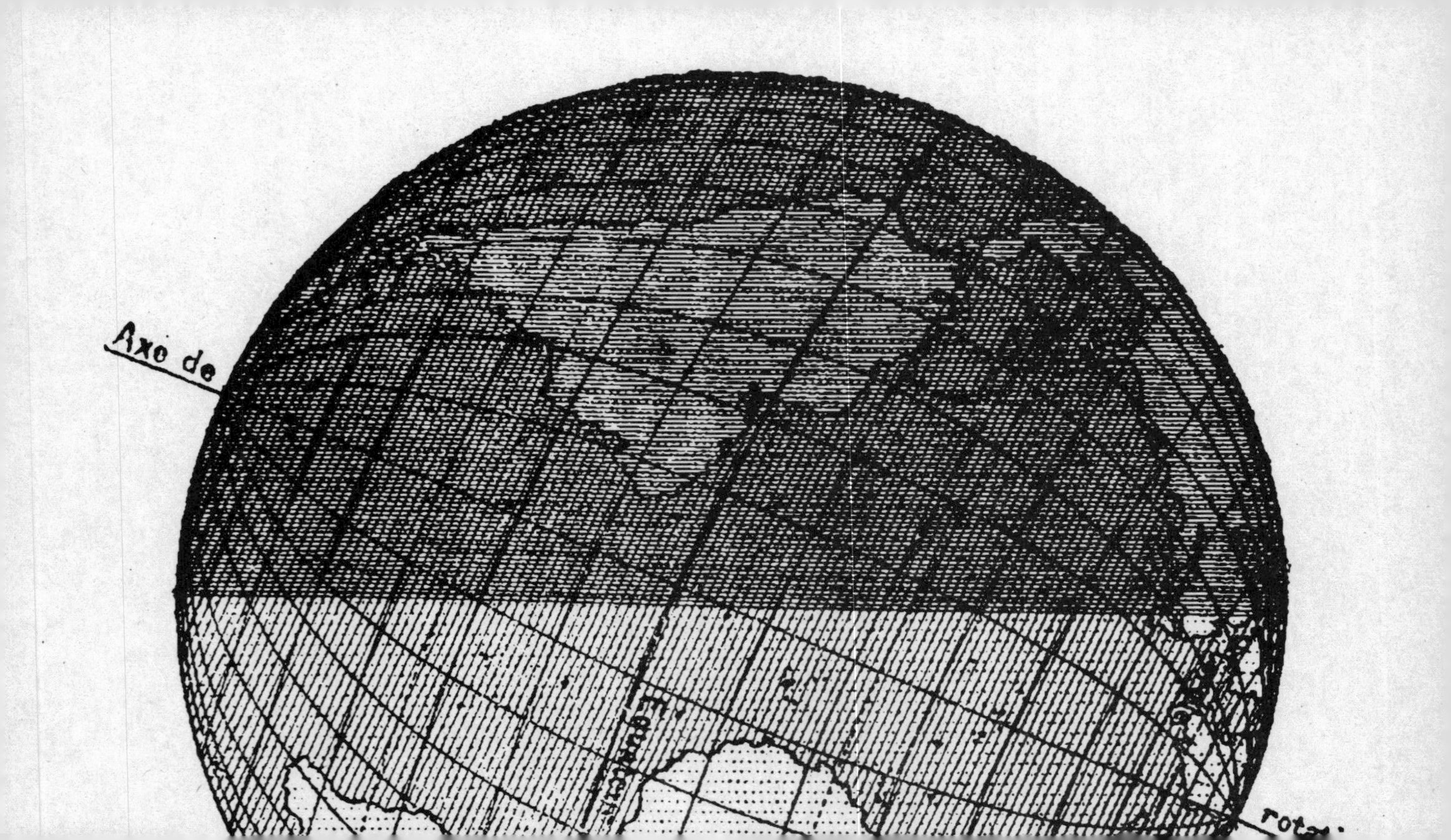
Axo de

PART TWO

OUR PLANET

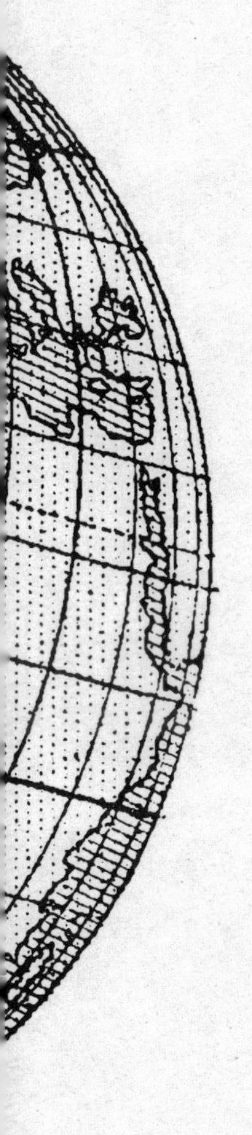

WHY ARE WE ON THE THIRD ROCK FROM THE SUN?

LOOKING AT THE EIGHT PLANETS OF OUR SOLAR SYSTEM, YOU'D BE HARD PRESSED TO SEE A FAMILY RESEMBLANCE. However, the story of the solar system's origin reveals that they were all created from the same raw material.

You might also think that these bodies are scattered across the solar system without rhyme or reason. But move any piece of the solar system today, or try to add anything more, and the whole construction would be thrown fatally out of kilter.

So how exactly did this delicate architecture come to be? The story begins 4.6 billion years ago, when something was brewing in a backwater of the Milky Way. The wispy stuff that permeates the in-between bits of all galaxies – hydrogen and helium gas, with a sprinkling of dust – had begun to condense. Unable to resist its own gravity, part of this cloud collapsed in on itself. In the ensuing heat and confusion, a star was born – our Sun.

Born from a dying star

We don't know what kickstarted this process. Perhaps it was the shock wave from the explosive death throes of a nearby star. But it was not a particularly unusual event. It had happened countless times since the Milky Way came into existence about 8 billion years before, and we can see it still going on in distant parts of our galaxy.

When the Sun formed, it swallowed about 99.8 per cent of the material in the cloud. The scanty remains were sculpted by gravity into a flattened disc encircling the new-born star. As the dust grains of this protoplanetary disc orbited the Sun, they collided and progressively coagulated into ever larger bodies called planetesimals.

Once they reached about a kilometre across, their gravitational pull was enough to start dragging in surrounding material, including other planetesimals, in a runaway process that eventually led to the formation of planets proper.

How this process played out depended on proximity to the Sun. The disc's innermost region was very hot, which meant only metals and minerals with high melting points were present in solid form. And so the planetesimals in that region could grow only so big. The result was the four small rocky planets of the inner solar system: Mercury, Venus, Earth and Mars.

Gas and ice

No such limitations existed further out, beyond the 'ice line' where methane and water were also present as solids. Here the planets could grow large enough to start collecting molecules of hydrogen and other gases. That was how the gas giants Jupiter and Saturn came to be, and, further out in still colder climes, the ice giants Uranus and Neptune.

So far, so straightforward. But when it comes to details, the model becomes rather vague. No one really knows how small boulders managed to coalesce into bodies thousands of kilometres across. Objects that small would have been buffeted around by the surrounding gas and sent spiralling into the Sun before they could stick together. Perhaps local patches of turbulence provided vortices of lower pressure in which the boulders could collect and coagulate.

A similar problem bedevils the gas giants. The risk of such planets being bounced towards the Sun is illustrated by the 'hot Jupiters' seen in other planetary systems. These are roughly the size of our Jupiter but are orbiting around their stars at the distance of Earth or closer. Had anything like that happened in the early solar system, the Earth and other inner planets could have been slingshotted out altogether.

The outer solar system does seem to have undergone a violent upheaval a few hundred million years after the Sun was born. Models suggest that the gas giants were closer together. Then something happened that made this arrangement unstable, hurling the planets to their present-day positions. Since then, however, the objects that constitute our solar system have settled into a tranquil, if sensitive, balance.

The solar system isn't just the Sun and planets, of course. Between Mars and Jupiter is a band of rubble called the asteroid belt, a ring of protoplanetary material that failed to accrete, possibly because of the gravitational influence of Jupiter. The belt is mostly full of rocks, but it does contain four quite large objects, Ceres, Vesta, Pallas and Hygiea, that together make up about half of its total mass.

Moons galore

The solar system is also full of moons; more than 180 have been named so far. Only three are in the inner solar system: ours, and two around Mars. The rest orbit the gas and ice giants. Most are thought to be either the remnants of the accretion disc, or passing asteroids that were captured by the planet's gravity. The rings of Saturn and Neptune may also be the remnants of the material that gave rise to the solar system, but their origin is not clear.

Beyond the ice giants lies the Kuiper belt, a frigid sea of perhaps 100,000 icy bodies including the former planet Pluto and its companion Charon. These objects are also remnants of the solar system's formation. The belt is huge, stretching from the orbit of Neptune – which is 30 astronomical units (AU), or 30 times the distance from the Sun to the Earth – almost as far out again, to 50 AU.

But this is still nowhere near the edge of the solar system. Further out still is the Oort Cloud, a sphere of mainly icy objects that extends out to about two light years from the centre. Here the Sun no longer wields any gravitational influence. The Oort Cloud has never been observed directly, but some astronomers suspect that it hides something that would radically change our view of the solar system: an ice-giant planet that we have not yet observed.

Who says there's nothing new under the Sun!

WHAT'S EATING JUPITER?

We call them gas giants, but most astronomers believe that Jupiter and Saturn have rocky cores. These formed in the same way as Earth, but once they reached about 10 times its mass, their gravity pulled in gas to create a thick atmosphere.

Curiously, some studies have suggested that Jupiter's core weighs less than it should. That may be because it is dissolving. The core endures enormous temperatures and pressures. Under these conditions the mineral magnesium oxide – thought to be a key constituent of the core – would dissolve into the atmosphere.

IS THE MOON REALLY A PLANET?

THERE ARE MORE THAN 180 OF THEM IN THE SOLAR SYSTEM, BUT OURS IS UNIQUE. Our moon – *the* Moon – may not be the largest out there. Unlike some of the icy moons of the outer solar system, it has no hope of harbouring life. It may be colder, quieter and more boringly spherical than some of its rivals. But when it comes to origins, no other moon has such a fascinating, turbulent story to tell.

Despite being only the fifth largest moon in the solar system – dwarfed by Saturn's Titan and Jupiter's trio of Ganymede, Callisto and Io – the Moon is still staggeringly huge. At over a quarter of the Earth's diameter, it is by far the largest moon in the solar system relative to the size of the planet it orbits.

The anomalous size of the Moon has made accounting for its origins a problem. Such a body could not have been formed like the other moons, which are either captured asteroids or remnants of the disc of dust and gas that became the solar system.

In 1879 George Darwin, the astronomer son of Charles, proposed a solution. He suggested that the Earth and Moon were once a single body that spun so quickly it fell apart, flinging a bit of itself into orbit. This molten rock eventually condensed into a blob that solidified to form the Moon.

This idea was popular for a time – the Pacific Ocean was once proposed as the scar left over from the event – but it fell out of favour in the twentieth century, when it turned out that the numbers didn't add up. For the young Earth to spit out a moon it would have had to be spinning at impossible speeds, rotating fully about once every two hours.

The Big Splat

As Darwin's idea fell out of favour another took its place. Known as the giant impact hypothesis or Big Splat, the idea is that at about 50 million years after the solar system began to form, a Mars-sized object named Theia collided with Earth. Striking our planet with a glancing blow, this body shattered on impact, sending up a plume of debris that eventually coalesced to become the Moon.

At first, there was nothing much to favour the Big Splat over any other explanation. It was proposed because nothing else worked. But that changed as astronomers refined their picture of what the early solar system was like. We now know that huge impacts were an important factor in planet formation.

The Big Splat is now the most widely accepted explanation, but it has problems of its own, which are leading some astronomers back to a modified version of Darwin's idea. According to the Big Splat, the majority of the Moon originated from Theia; Earth contributed only a small amount of material. If so, you'd expect Moon rocks and Earth rocks to be substantially different in composition (unless, by some other giant coincidence, Theia and Earth were made of exactly the same stuff). But that is not what analyses of Moon rocks show.

Watery moon

In fact, the Moon's oxygen, chromium, potassium and silicon composition is indistinguishable from that of Earth. Moon rocks also contain surprisingly large amounts of water. In the hellish aftermath of a giant impact, the heat generated should have driven off the water.

The findings suggest instead that the Moon was once a part of Earth that was somehow blasted into space without being contaminated by a colliding planet. To avoid the problem that plagued Darwin's solution, however, requires a massive input of energy from somewhere.

One dramatic but still very speculative idea is that this energy – equivalent to 40 million billion Hiroshima bombs – came from within, in the form of a giant nuclear explosion.

This is not as outlandish as it first seems. We know that the Earth once contained naturally occurring concentrations of uranium, which would have behaved like nuclear reactors. Their dormant remains have been found in numerous locations around the world, most notably in Gabon.

These reactors were active around 2 billion years ago and probably burned for a few hundred thousand years until they exhausted their supply of uranium. But they were no more than about 10 metres across, far too small to blow the Earth apart.

But something similar could still explain the origin of the Moon. The basic idea is that radioactive elements such as uranium, thorium and plutonium were concentrated in dense rocks that sank deep into Earth shortly after its formation. They accumulated at the boundary of the outer core and the mantle, where geological forces brought them closer together to form a colossal nuclear reactor that eventually went supercritical and exploded with enough force to eject a moon-sized glob of rock into orbit.

A smaller splat?

There are other ways to get around the problems of the Big Splat. One is that a body about half the size of Theia struck full on and buried itself deep within the Earth. Computer simulations show that this would provide enough energy to explode a plume of molten material into orbit, creating a moon made of rocks indistinguishable from those on Earth.

A different sort of 'small splat' envisages two planets, each about half the size of Earth, colliding slowly. The ensuing coalescence gave birth to our planet and formed the Moon from the leftovers.

The good news is that, even if the nuclear option is correct, there's no risk of a repeat performance. The radioisotopes that powered the explosion have mostly decayed away by now. When it comes to another catastrophic impact, we may not be so lucky.

DARKNESS AT NOON

If you've witnessed a total solar eclipse, you have experienced one of the most astonishing coincidences in the solar system. As the Moon slides over the face of the Sun it fits perfectly, obscuring the Sun completely but leaving its corona visible as a glowing halo. That spectacle is possible because the Moon and Sun are almost exactly the same size in the sky (the Sun's diameter is actually 400 times larger than the Moon's, but it also happens to be about 400 times further away).

This seems like it should be significant, but is actually a coincidence. The Moon was once much closer to the Earth, and is creeping away at about 3.78 centimetres a year. So total solar eclipses of the deep past and distant future would have been and will be much less impressive.

WHY DOES EARTH HAVE LAND AND OCEANS?

THE EARTH IS ONE OF FOUR ROCKY PLANETS IN THE SOLAR SYSTEM, ALONG WITH MARS, VENUS AND MERCURY. But in many respects it doesn't belong – and that's not just because it harbours life. Earth has a restless surface that is constantly rearranging itself through the slow but inexorable process of plate tectonics. Even more unusually, about 70 per cent of its surface is covered in water.

'How inappropriate', Arthur C. Clarke once said, 'to call this planet Earth when it is quite clearly *Ocean*.' To understand how Earth acquired its shifting, watery surface, we need to go back to its earliest origins.

Time zero for the solar system is generally agreed to be 4.567 billion years ago. At that point Earth did not really exist; it was still in the process of being built from violent collisions between planetesimals. By 4.55 billion years ago about 65 per cent of the Earth had assembled. This embryonic planet was very hot, a 'magma world' composed entirely of molten rock. But it presumably began to cool and form a rocky crust.

Then, 20 million or so years later, as the infant Earth was

solidifying and settling down in its orbit around the Sun, it was rudely interrupted. The young planet was dealt a glancing blow by an object the size of Mars. Debris from the impact was thrown into Earth's orbit and eventually formed the Moon. The planet's atmosphere filled with vaporised rock, which condensed and fell as lava rain, depositing a sea of molten rock at a rate of perhaps a metre per day.

Earth takes a pounding

The energy of the collision also delivered enough heat to melt the Earth once more, perhaps right down to its core, recreating the magma ocean and erasing our planet's previous geological record.

Then, as if to add insult to injury, about 4.1 billion years ago the Late Heavy Bombardment started. Possibly triggered by an orbital rearrangement of the gas giants, meteorites from the asteroid belt rained down on Earth, partially remelting its surface all over again. These cataclysms, combined with later plate tectonics and weathering, have left a yawning chasm in our knowledge of the Earth's first half-billion years, an era known as the Hadean after the hellish conditions that prevailed.

When Earth emerged from the Hadean, 4 billion years ago, it had dry land, oceans, plate tectonics and, possibly, life. It wasn't exactly like the planet of today – the atmosphere was very different, for example – but it was a far cry from the ball of molten rock that emerged from the Moon collision. How did it get that way?

The Earth's crust today is composed almost exclusively of rocks no older than 3.6 billion years, so traces of the Hadean environment are extremely thin on the ground. Of the truly ancient rocks that remain – amounting to about one part per million of the crust – most have since been melted and compressed, then remelted and recompressed, beyond recognition. But thanks to tiny resilient crystals called zircons there are some clues as to what happened.

As old as the hills

Found mostly in the Jack Hills of Western Australia, zircons are Earth's oldest minerals. They are composed of exceptionally durable zirconium silicate crystals and contain a high concentration of uranium, which allows them to be dated from the amount of remaining radioactivity. Many zircons date to more than 4 billion years old, meaning they formed in the Hadean.

They cannot tell us exactly what happened as the battered Earth recooled, but their oxygen content shows that they formed in water, suggesting that Earth's oceans were in place more than 4 billion years ago. This raises many new questions, not least where the water came from and how it managed to not simply boil off. It also tells us that Earth must have acquired a crust during the Hadean, as oceans need to sit on a solid surface.

The first crust was basalt, a dense, black volcanic rock that still forms at mid-ocean ridges. Basalt makes up much of the ocean floor and can accumulate in such large volumes that it emerges as dry land from the middle of the ocean – Iceland and Hawaii are both massive lumps of basalt that erupted from the sea floor. This process also occurred on early Earth. For the first billion years or so, the surface of the Earth consisted of seas interrupted by chains of volcanic islands.

World in motion

The first continents also began to form around this time. They were made from another volcanic rock called granite. This is a lighter rock which forms at subduction zones – regions where slabs of ocean floor slide under one another or beneath continents. Granite is much less dense than basalt, and so it floated on the denser rock. This is how the first true continental crust formed. Some bits of continental crust that formed 4 billion years ago still exist today.

The presence of granite tells us that plate tectonics must have got moving around this time too, though nothing like the full-blown (though glacially slow) churning of the Earth's surface today.

There are even suggestions that life started in the Hadean. The oldest reliable fossils are 3.43 billion years old, but there are chemical signatures in 4.1-billion-year-old zircon that may be the remains of living organisms. If so, how they survived in hell is a new mystery to solve.

COSMIC COCKTAIL

Earth had oceans very early in its history, but exactly how it acquired them is a mystery. Being so close to the Sun, it seems unlikely that it retained any water from the protoplanetary building material.

The standard explanation is that the oceans arrived during the Late Heavy Bombardment, as the payload of icy bodies such as comets and asteroids. The recent Rosetta mission to comet 67P Churyumov-Gerasimenko has cast some doubt on this idea, however. This comet's water does not have the same composition as Earth's.

The water on other comets and asteroids turns out to vary enormously. It appears that Earth's water is a mixture of different sources from across the solar system, which arrived in a life-giving bombardment more than 4 billion years ago. Think of that, next time you boil the kettle.

WHY IS THE WEATHER ALWAYS CHANGING?

If you picked a random location on the surface of the Earth and stood there for a year, you'd experience pretty much every sort of weather going. If you picked Trafalgar Square in London, for example, you could look forward to 10 days of torrential rain, 50 frosty mornings, 5 days of snow, about 15 thunderstorms, a gale or two, 1,500 hours of sunshine and an awful lot of cloud.

Why is it that the same spot on Earth can experience such diverse weather? The answer is above our heads, in a thin layer of gas enveloping the Earth and a huge ball of hot gas 150 million kilometres away.

Weather from the Sun

Whether the weather be hot, or whether the weather be cold, the cause is always the same: radiation from the Sun hitting a roughly spherical, spinning planet with a gaseous atmosphere. Because of this simple set-up, the atmosphere is heated unevenly. At the equator, sunlight strikes the Earth from directly overhead; at the poles it arrives at a sharply slanted angle. The polar regions thus receive less sunlight for a given area than the equator does. That is why the poles are cold and the equator is hot.

From this difference, weather flows. Heat naturally moves from hotter to colder areas, so the atmosphere and oceans transport heat from the equator to the poles. A planet without temperature differences would be a planet without weather.

If that was all there was to it, global weather patterns would be very simple. Hot air would rise at the equator and move towards the poles. There it would cool, sink and flow along the surface back to the equator. Surface winds would thus flow uniformly from the tropics to the poles.

But that is not what happens, for the simple reason that the Earth rotates. On a rotating sphere, the surface – and the air above it – moves fastest at the equator and not at all at the poles. Thus, Earth's rotation deflects the north–south winds to the side. This deflection is called the Coriolis effect.

The rotation of the Earth creates a Coriolis effect strong enough to disrupt the basic north–south flow and create six interlocking bands of surface winds, three in each hemisphere: the polar easterlies, the mid-latitude westerlies and the equatorial trade winds. Where the trade winds meet is a band of erratic weather called the Intertropical Convergence Zone.

Coriolis forces also create winds far above the surface of the Earth. These fast-moving west–east bands of wind are called jet streams. Earth has four, two in each hemisphere: a polar one and a subtropical one.

This basic pattern predominates, but actual winds are more

complicated and variable. This is because Earth is not a uniform sphere but one with oceans, mountains, forests and deserts, all of which influence the movement of the air.

Big fluffy clouds

Besides wind, the other fundamental ingredient of the weather is water, which we experience in the form of clouds and rain.

Two things need to be present to create clouds: water vapour in the air and a mechanism to lift it upwards. Water vapour gets into the air from evaporation of surface water and transpiration by plants, which suck water out of the soil and release it through their leaves. The lifting mechanism is supplied in three ways. The first is rising parcels of warm air known as thermals. The second is when air masses of different densities meet and create fronts that push air upwards. The third is air being blown against mountain ranges and forced upwards.

As air rises, it cools and expands. At some point it becomes too cold for water vapour to remain in the gas phase. When air reaches this dew-point temperature, water begins condensing out, forming clusters of tiny droplets – i.e. clouds. If the droplets become large enough, they fall out of the sky as rain, sleet, snow or hailstones.

All this happens in the lowest 7 to 20 kilometres of the atmosphere, otherwise known as the troposphere. Above this altitude the air suddenly begins to warm again due to ozone absorbing ultraviolet light. This is the lower boundary of the stratosphere. And that is it. These factors are enough to explain all the weather events we experience, from calm balmy days to violent storms.

Thunder, lightning, strike!

Among the most violent of all are thunderstorms. If the Sun's heat is strong enough, thermals create cauliflower-shaped cumulus clouds, which can reach the top of the troposphere. Freezing temperatures in the upper reaches create ice crystals, and collisions between them separate electric charge. When the separation builds up to a critical level, the charges reunite in a lightning bolt. The cloud is now a thunderstorm – though the cause of the thunderclap itself is still unclear.

The world's heaviest rainfalls are invariably caused by thunderstorms. And when wind conditions are just right (or just wrong, depending how you look at it) thunderstorms also give rise to nature's most violent windstorms, tornadoes. During a tornado in Oklahoma in May 1999, radars clocked a wind speed of 486 kilometres – the fastest ever recorded.

Tropical cyclones – which include hurricanes and typhoons – are another extreme weather system. Though less violent than tornadoes, they are absolutely huge – up to 2,000 kilometres across, capable of generating storm surges of over 10 metres and dumping more than a metre of rain a day.

Tropical cyclones form over the ocean where the sea surface temperature exceeds 27 degrees Celsius, producing a large amount of evaporation. When this water vapour condenses, the release of latent heat cooks up a tropical storm. If the storm is set rotating by a combination of wind and the Coriolis forces, the result can be the most destructive weather system on Earth.

A HAPPY TEDIUM

Extreme weather grabs headlines, and the changeable weather of the British Isles gives their inhabitants something to talk about. But have you ever wondered where on Earth has the least extreme, least changeable weather? *Weatherwise* magazine tried to find out. It crowned Viña del Mar, a coastal town near Valparaiso, Chile, as the place with the most unexciting climate. The daytime temperature wavers between 15 and 25 Celsius all year round. It is usually slightly overcast, and drizzles quite a lot. The wind rarely gets above a stiff breeze. It never freezes or snows. Only the occasional thunderstorm breaks the tedium.

WHERE DOES SOIL COME FROM?

COMMON AS MUCK. Dirt cheap. Soiled. The earth beneath our feet rarely moves anyone to poetry. But look at it closely, and you'll see it is a thing of beauty.

Soil covers much of the surface of the Earth. Without it, our planet would be a very different – and extremely hostile – place.

Soils vary enormously, but roughly speaking they are a 50/50 mixture of solid material and holes. The solid stuff is largely small pieces of rock plus organic matter, both living and dead. The holes are not empty space, but are filled with water and gas in variable proportions. This simple list of ingredients does not constitute soil, however. To get the finished product, everything needs to be cooked up in a complex and very lengthy recipe.

The starting point for most soils is bare bedrock. This is eroded by weathering, generating ever-smaller fragments that accumulate on the surface. 'Weathering' is an apt word. It occurs when the rock is battered by wind, rain and hail, cycles of freezing and thawing further weakening and shattering it. Thermal expansion and contraction as the temperature rises and falls has a similar effect.

Rocks are also worn down by chemicals in rainwater, which dissolve certain minerals. Finally there is biological weathering. Bare rock is initially colonised by bacteria and other microbes, which excrete corrosive acids. Lichens and algae come next; their physical attachment to the rock is a powerful agent of erosion. Experiments on barren land in Hawaii suggest that lichens speed up weathering by at least 100 times.

In mature soils, biological weathering is even greater. Respiration by invertebrates, fungi and bacteria pumps out carbon dioxide, which accumulates between soil particles. Rainwater percolating through the soil dissolves the CO_2 to form carbonic acid. Other acids are produced by soil organisms. Soil also acts like a sponge, prolonging the time that the underlying rock remains wet after rain, which means chemical weathering can happen for longer. In this way, soil is a catalyst of its own production.

The initial microbial colonisers also get the ball rolling on the organic matter in soil. The material they leave behind is exploited by lichens and algae, which gradually cover the rock in living, and then dead, organic matter. Once enough of this has built up, larger organisms such as worms and arthropods move in. Their burrowing mixes the organic matter and mineral grains together and creates pore spaces. Mucus produced by worms also glues the matter together and stabilises it. A soil is born.

Growing old gracefully

As the soil deepens and matures it may differentiate into layers, with topsoil on the top and various subsoils beneath. A mature soil is absolutely teeming with life. A single gram may contain 100 million individual bacteria and archaea, 10 million viruses and 1,000 fungi, not to mention the larger organisms and plant roots. The bacteria alone may represent a million different species.

Unsurprisingly, this all takes time. Weathering is a laborious process. Lichens grow painfully slowly. Research on recent lava flows in Hawaii suggests that it takes at least a century and as much as 10 millennia to create even rudimentary soil. Lava flows that formed 100 years ago remain almost totally barren; even those that are 10,000 years old have only just developed something resembling soil. This suggests that the rich soils covering much of the Earth have taken thousands of years to form. Some soils in Africa and Australia have been dated back to 144 million years old, during the Cretaceous period.

And soils go way further back than that. The earliest known palaeosols – fossil soils – date back more than 2 billion years, long before plants even existed let alone colonised the land. Far from being primitive, these soils are thick and well developed. Some contain as much as 50 per cent clay minerals by volume, the end products of extensive bedrock weathering. This is characteristic of a soil that has been stable for hundreds of thousands of years at least.

Peak soil

There is no reason to think that palaeosols formed through a process that was radically different to modern soil, though there were no multicellular organisms – plants, worms and arthropods – to work their magic. Lichens were probably also absent, though their fossil record is too sparse to know for sure. The most plausible scenario is that the soil formed through the action of hardy bacteria that colonised the land surface billions of years ago.

The closest analogue we have to this lost world is found in Canyonlands National Park, in the high desert of Utah. Under a harsh sun and bracing winds, bacteria, lichens and mosses eke out a precarious existence on the surface of rocks. Together they form what is known as a cryptogamic crust on the surface of the rock. Intermingled with this crust is a thin layer of mineral and organic debris – in other words, a soil. Visitors are asked to protect the fragile soil by sticking to marked trails; even a single footstep can break the crust and expose the soil beneath to erosion. Once erosion starts, it can spread catastrophically.

And protection is something we need to do on a global scale. According to the United Nations, more than a third of the world's soil is endangered by agriculture and construction and we are losing fertile topsoil at a rate of 30 soccer pitches a minute. Given that soil grows 95 per cent of our food, holds three times the amount of carbon as does the entire atmosphere and takes thousands of years to replace, we really need to take action.

Save our soils!

20,000 SHADES OF BROWN

Soils today are an incredibly diverse bunch. The bedrock, the climate, the terrain, the local ecosystem and the age of the soil all influence its composition. This diversity is captured by a classification system every bit as elaborate as the one we use to categorise life forms. The US Department of Agriculture, for example, breaks it down into orders, suborders, great groups, subgroups, families and finally series – the equivalent of a species. In the US alone, more than 20,000 soils have been catalogued.

WHY DOES EARTH HAVE SUCH A GREAT ATMOSPHERE?

TAKE A DEEP BREATH. You have just inhaled around 26 thousand billion billion molecules of gas, mostly nitrogen and oxygen. But if you could have taken a breath on the surface of the early Earth, you'd have sucked in a very different 26 thousand billion billion molecules, mostly carbon dioxide and sulphur dioxide. (You wouldn't be taking very many more breaths after that.) Our air might be largely out of sight and out of mind, but its presence is one of the miracles that sets the Earth apart from any other planet we know of.

Earth's atmosphere is today made up of about 78 per cent nitrogen by volume, 21 per cent oxygen, 1 per cent argon and variable quantities of water vapour. There are also small traces of carbon dioxide, sulphur dioxide, carbon monoxide, methane, helium, neon and krypton, and even smaller traces of ozone, hydrogen, xenon, radon, oxides of nitrogen and man-made industrial pollutants such as chlorofluorocarbons.

The composition has changed radically since the atmosphere first formed. Very early on, Earth was probably surrounded by a tenuous atmosphere of gas – mostly hydrogen – left over from our planet's formation. But this 'first atmosphere' didn't last long, swept into space by gusts of solar wind. So we can rule this out as the origins of today's air.

Earth quickly acquired a second atmosphere from an unlikely source: its own innards. Volcanoes expelled heavier gases, which, under the pull of Earth's gravity, were unable to escape out into space. Comet and asteroid impacts could also have added some gases to the atmosphere. So our present-day atmosphere has evolved from a mixture of Earth farts and space burps.

The second atmosphere would have been dense and choking, made up mostly of steam, carbon dioxide and sulphur dioxide. We know this because these are the main gases emitted by volcanoes today. Vulcanism was extremely active and atmospheric pressure was probably 10 times its current level – which explains why the early oceans did not boil off into space.

Oxygen also began to accumulate slowly at this time, as sunlight fractured molecules such as carbon dioxide and water. But it remained an insignificant component of the atmosphere until much later.

So how did the mixture of carbon and sulphur dioxides, belching and hissing from volcanoes over billions of years, evolve into an atmosphere that is mainly nitrogen and oxygen? There are two answers. First, great quantities of carbon dioxide dissolved in the oceans and were eventually laid down as limestone. Second, life emerged and radically altered the composition of the atmosphere.

Primordial smog

At first, life's main contribution to the atmosphere was methane, a waste product of primitive single-celled organisms liberating energy from hydrogen and carbon dioxide. Around 3.7 billion years ago a 'methane crisis' nearly wiped life off the face of the Earth almost as soon as it had got going. Methane-belching microbes filled the atmosphere with a smog that all but blocked out the Sun.

The next big change was the Great Oxygenation Event around 2.3 billion years ago. The seeds for this cataclysm were sown around a billion years earlier when certain microbes evolved a new way to liberate energy from sunlight, called photosynthesis. One of its waste products was a highly toxic and violently reactive gas that had scarcely been seen on Earth before – oxygen.

The first photosynthesisers did not dump their poisonous waste product directly into the air, but locked it safely away in iron compounds. The result was the production of layers of iron oxides known as banded iron formations, which are found around the world in rocks between about 3 billion and 1.5 billion years old.

But then new photosynthetic organisms evolved that were able to tolerate free oxygen. They released their toxic waste directly into the air, saving the effort required to lock it away and, as a bonus, killing off many competitors. Oxygen began to accumulate in the atmosphere, rising from about 1 per cent to 10 per cent or more.

The Great Oxygenation Event is also called the Oxygen Catastrophe because it almost poisoned life out of existence. But evolution saved the day by inventing a way to utilise oxygen, called respiration.

The Oxygenation Event precipitated another catastrophe. Photosynthesis sucked the greenhouse gas CO_2 out of the atmosphere and ultimately locked it away in sedimentary rock. Meanwhile, oxygen reacted with methane, an even more potent greenhouse gas. Together these tipped the world into a global ice age called Snowball Earth, which lasted about 400 million years until a huge pulse of vulcanism refilled the atmosphere with greenhouse gases. The snowball also seems to have dragged oxygen back down to very low levels, perhaps because photosynthesis almost stopped. But as the ice melted and life rebounded, the Oxygenation Event happened all over again.

Breath of life

It was not all bad news, however. Oxygenation eventually helped to keep the planet habitable by forming the protective ozone layer about a billion years ago. While the drama was unfolding, inert nitrogen continued to leak out from volcanoes. And because it had nothing to do and nowhere else to go, it gradually built up to become the most abundant gas in the atmosphere. By about 600 million years ago the composition of the atmosphere was roughly what we are familiar with today.

Its composition and density has varied with time, driven by a complex interplay of biological, geological and chemical processes. For example, about 300 million years ago oxygen peaked at about 30 per cent, allowing flying insects a metre long to evolve.

Nonetheless, for the past half-billion years the air has been essentially the same as the stuff you are breathing right now.

ALIEN AIR

If you want a rocky planet with atmosphere, Earth is about as good as it gets. Mars has almost none – just a trace of carbon dioxide at a pressure less than 1 per cent of Earth's. This is largely because Mars is smaller than Earth so its gravitational pull is not strong enough to retain a layer of gas. Mercury is even smaller and even more lacking in atmosphere. Venus, however, has swung the other way: it is swathed in hot, dense clouds of volcanic gases and sulphuric acid at a pressure almost 100 times that on Earth. But 70 km above its surface, the atmosphere is downright balmy – plenty of sunshine and water, and Earth-like pressures and temperatures – and may be just right for life.

HOW DID OUR PLANET FILL UP WITH PETROL?

NEXT TIME YOU TRAVEL IN A CAR, BUS OR TRAIN, CONSIDER THIS: THE STUFF THAT IS FUELLING YOUR JOURNEY IS FOSSILISED SUNSHINE THAT HASN'T SEEN THE LIGHT OF DAY FOR TENS OR EVEN HUNDREDS OF MILLIONS OF YEARS.

Oil is the lifeblood of modern civilisation. It has become critical to our prosperity and security. Wars have been fought over it, and we have no coherent plan for how to live without it once it is gone. Every day we use nearly 90 million barrels of the stuff – enough to fill London's O2 Arena five times over.

Plankton power

All of this is very grandiose considering that the vast majority of the world's oil started life as plankton floating in an ancient ocean quietly converting sunlight into organic molecules via photosynthesis. When they died, their bodies sank to the ocean floor where oxygen was too scarce for decomposition to occur.

Their energy-rich remains accumulated into a thick organic sludge, mixed with silt, sand and other inorganic matter, and gradually disappeared under layers of sediment.

Over millions of years the sludge was buried ever deeper as more sediment piled on top. Once it reached a depth of 3 kilometres, heat from below and pressure from above started to cook the organic molecules, breaking them up into simpler hydrocarbon chains. The first product was a waxy solid called kerogen. This was further broken down or 'cracked' to form a mixture of liquid hydrocarbons called petroleum, plus methane or natural gas. Sometimes the temperature was too high and all the organics were broken down to methane. That 'overcooking' usually happens if the deposit is more than 5 kilometres down.

But when conditions are just right, petroleum forms. The composition of the finished product depends on the starting material and what combination of heat and pressure it has been subjected to. Low-temperature petroleums are thick and black, like tar; higher-temperature ones are thin and clear, like gasoline. The colour can range from black to brown, green or even yellow. And the proportion of the really valuable compounds – the paraffins that get processed into fuel – can vary from as little as 15 per cent to as much as 60 per cent.

That is not the end of the story. Petroleum rarely pools in underground reservoirs. It is integrated into the rock – often along with water – and has to be separated out. What is more, it only forms exploitable reserves under certain circumstances. The rock in which it formed has to be porous, so the liquid and gas can move through it, rising towards the surface. There also has to be a trap – perhaps a layer of dense, non-porous rock, or a fault – above the oil-bearing rock to stop it from seeping to the surface. The cap rock also has to be the right shape to allow oil and gas to accumulate underneath it. Only then will oil and gas form the large deposits that we call oil fields.

Well, well, well

Fortunately – or unfortunately, depending on how you look at it – petroleum is abundant and the Earth is very large, and so the geological stars align often enough for exploitable oil and gas deposits to be quite common. There are something like 65,000 known oil and gas fields, and geologists continue to discover new ones. The stuff that comes out of these wells is called crude oil, and is the starting point for a range of products including the petrol you put in your tank and many of the plastics that define the modern world.

Determining the precise origin of petroleum is quite difficult, as it often migrates large distances underground and can't be dated by the rocks it is found in or under. But knowing when it formed can help oil geologists understand what is beneath their feet, and hence where they should concentrate their exploration efforts.

Oil is generally dated using biomarkers, organic compounds that are characteristic of different eras of life. A compound called oleanane, for example, is made only by flowering plants, so oils that contain it must date from the Cretaceous period or later (pollen is a small but not insignificant contributor to the organic matter that eventually becomes oil).

Biomarker analysis has shown that some oils are very old indeed, dating from before complex life evolved 540 million years ago. Others are quite recent, as little as 5 million years. Petroleum is generally thought to require several million years to form, though some very young deposits have been found, which suggests that this isn't always the case. Mature oil as young as 5,000 years old has been found in the Gulf of California, and Russian geologists claim to have found oil in Kamchatka that is only 50 years old.

Some oil may be non-biological in origin, formed from carbon that was present when the Earth formed or brought in by comets, but even if such oil exists, it forms a negligible proportion of our deposits.

From beneath a vanished ocean

If there is a golden age of oil formation, it is probably the Jurassic period, from 200 to 145 million years ago. Oil formed in massive quantities at the bottom of the Tethys Ocean which once separated the palaeocontinents Gondwana and Laurasia. The ocean eventually closed due to continental drift, though remnants still exist – the Mediterranean, Black, Caspian and Aral seas are fragments of the Tethys Ocean. But its most important legacy is the vast stores of energy now found under the half-dozen or so Middle Eastern states that supply two-thirds of the world's oil.

North Sea oil also formed during the Jurassic period. So chances are, the energy that drives your car was extracted from sunlight by a plankton that died 200 million years ago.

FROM SLUDGE TO BLACK GOLD

The world's first oil well was drilled near Titusville, Pennsylvania, in 1859, an area known as Oil Creek because of the bituminous substance that bubbled up from the ground. Residents used it as a medicine, but the Pennsylvania Rock Oil Company had better ideas: they wanted to develop a brand new industry.

Their product was light. The company realised that 'rock oil' contained kerosene, an excellent illuminant for oil lamps. Kerosene quickly became big business.

At that time gasoline was an almost useless by-product that was often just thrown away. But at the turn of the century, just as Thomas Edison's light bulb was killing off the kerosene business, the oilmen hit paydirt. Henry Ford made the internal combustion engine an essential feature of modern life and gasoline suddenly found a large and expanding market.

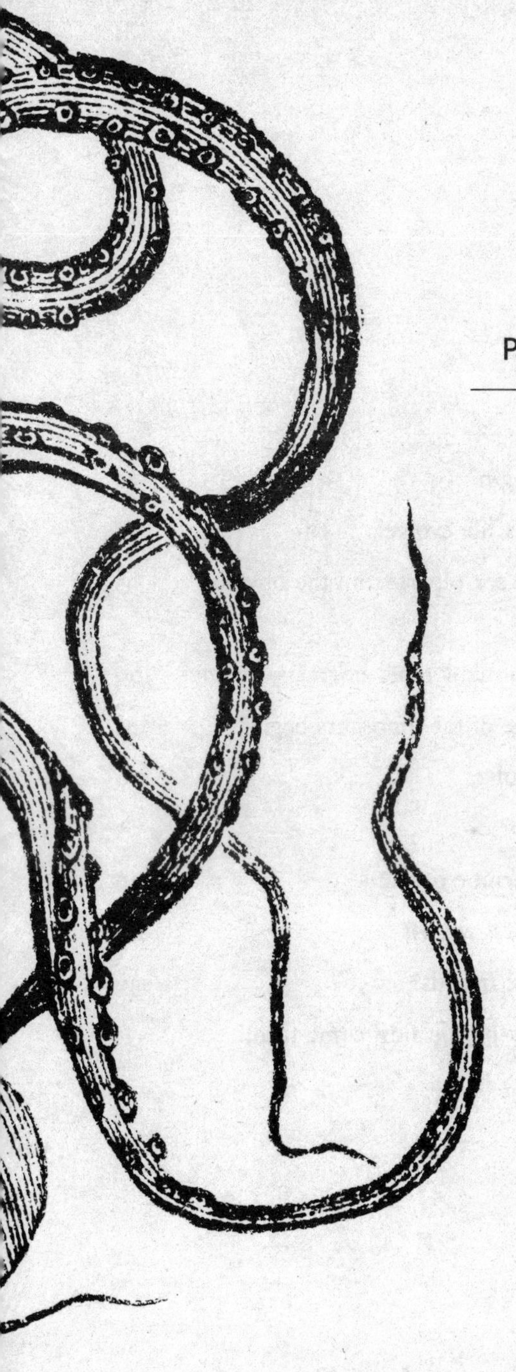

PART THREE

LIFE

WHEN DID LIFE BEGIN?

FOUR BILLION YEARS AGO: THE SURFACE OF THE EARTH IS BEGINNING TO COOL DOWN. It's a violent place bombarded by meteorites, riven by volcanic eruptions and enveloped by a toxic atmosphere. But despite the hostile conditions something extraordinary happens. A molecule, or perhaps a set of molecules, capable of replicating itself arises. Of all the amazing events our young planet has already witnessed, this is the most amazing of all.

Once the replicators appear, natural selection kicks in, favouring any offspring with variations that make them better at replicating themselves. Soon the first simple cells appear. Life has begun.

Darwin was among the first to speculate about how it happened. He envisaged a 'warm little pond, with all sorts of ammonia and phosphoric salts, lights, heat, electricity, etc. present'.

Warm ponds are no longer seen as a viable cradle of life. But others have been proposed, including the open ocean, deep-sea vents, radioactive beaches and lumps of clay. The bottom line is, we don't know where or how life started. But we know enough to make an educated guess.

The start of something big

The earliest undisputed fossil bacteria date from about 3 billion years ago, but it is widely accepted that life must have started much earlier, at least 3.5 billion years ago. Exactly when, though, is very hard to say. Many ancient rocks contain physical structures and chemical signatures that have been claimed as evidence of life; the earliest dates back to 4.1 billion years ago, though this seems a stretch given that Earth was still being pummelled by the Late Heavy Bombardment. Perhaps the best guess is 3.8 billion years ago.

If the 'when' is hard to pin down, the 'how' is even harder. Any theory of the origin of life needs to explain three things: how the building blocks assembled into complex molecules; how these were contained in a confined space like a cell; and where energy came from to drive the process. Perhaps the closest we have to a theory that explains all three centres on the sea floor, at places called alkaline hydrothermal vents. These are distinct from the well-known volcanic hydrothermal vents, or 'black smokers', where superheated water gushes from volcanic crevices.

Alkaline vents, which are found on Earth today and are presumed to have been common on the early Earth, are much less turbulent. They are fissures in the seabed that gently seep lukewarm alkaline fluids.

These vents are formed when seawater percolates into the seabed and reacts with a mineral called olivine. This reaction enriches the water in hydrogen and generates heat, which drives the fluid back up to the ocean floor. As the warm fluid hits the cold seawater, minerals precipitate out, gradually forming delicate rocky chimneys up to 60 metres tall. Such structures provided everything needed to incubate life.

Building blocks

First, chemicals. The chimney walls would have been rich in minerals which catalyse the formation of complex organic compounds from CO_2 and hydrogen, both abundant in the vent fluids. This would have resulted in the spontaneous formation of the molecules that are the building blocks of life, including amino acids, sugars and, crucially, RNA.

RNA, a close cousin of DNA, is absolutely central to our ideas about the origin of life. When biologists first started to ponder the question, it seemed baffling. All living organisms rely on proteins to do their hard work. Proteins can fold into a wide diversity of shapes, so can do just about anything, including catalysing the chemical reactions of life. However, the information needed to make proteins is stored in DNA. You can't make new proteins without DNA, and you can't make new DNA without proteins. So which came first?

The discovery that RNA could both fold a bit like a protein and also catalyse reactions solved this chicken-and-egg problem. About 25 years ago it led to the idea that the first life consisted of RNA molecules that catalysed their own production. Alkaline vents seem like an ideal place for this RNA world to evolve.

Next, containment, to keep the molecules from diffusing apart. The vent itself could have done the job. Inside its porous structure were tiny, interconnected cell-like spaces enclosed by flimsy mineral walls. These could have contained and concentrated the RNA and other complex molecules forming on their surfaces.

The RNA world also needed energy, and again hydrothermal vents could have supplied it, in the form of a natural 'battery' where the fluid meets the seawater. The seawater is acidic (proton-rich) and the vent fluid alkaline (proton-poor), so where the two meet there is a steep difference in the concentration of protons. Because protons carry a positive charge, this gradient creates an electrical potential across the interface.

This energy would have further driven the reactions between CO_2 and hydrogen, accelerating the formation of complex molecules and longer strands of RNA. At some point, proto-cells evolved a way to exploit the gradient. One of the best pieces of evidence for this crucial step in evolution is that living cells are still powered by proton gradients across cell membranes.

Simple recipe

There are many steps to fill in. But alkaline hydrothermal vents were a perfect setting for the RNA world. They are not the only option on the table, but they are our best guess at the cradle of life.

Many other questions remain. How did life break free from the vents? How did it make the transition from RNA to DNA and proteins?

We may never know. But if the hydrothermal theory holds up, it tells us something quite profound. Far from being an unfathomable mystery, the emergence of life is an almost inevitable consequence of a planetary system with three basic ingredients: rock, seawater and carbon dioxide.

IT CAME FROM OUTER SPACE

One highly speculative idea about the origin of life on Earth is panspermia, which proposes that it arose somewhere else in the galaxy, possibly Mars, and was brought to Earth on a comet or meteor. If so, we are all aliens, and life must be older than the 4 billion years or so life has existed on Earth. However, panspermia doesn't answer the fundamental question of how and when life got going, it just moves it elsewhere.

HOW DID COMPLEX LIFE EVOLVE?

OUR LIVING PLANET, AS DARWIN WROTE IN *ON THE ORIGIN OF SPECIES*, IS ONE OF 'ENDLESS FORMS MOST BEAUTIFUL'. But go back a couple of billion years and things would have looked very different. Despite the fact that Earth had been alive for the best part of 2 billion years, life remained extremely rudimentary – just bacteria and their superficially similar but actually very different sister domain, archaea. The most complex living things were colonies of microbes such as stromatolites and microbial mats. No plants, no animals, just a barren landscape of rock, river and ocean.

The emergence of those endless forms most beautiful is arguably the most important event to have happened on Earth since life itself got going. It certainly seems to have been one of the most unlikely.

For many years, biologists assumed that the emergence of complex life was an evolutionary inevitability. Once simple life emerged, it gradually evolved into more complex forms, eventually giving rise to animals and plants. But that doesn't seem to be what happened. After simple cells first appeared, there was an extraordinarily long hiatus – nearly half the lifetime of the planet – before complex ones evolved. In fact, it appears that simple cells gave rise to complex ones just once in 4 billion years of evolution, which is suggestive of a freak accident.

Missing links

If simple cells had slowly evolved into more complex ones over billions of years, all kinds of intermediate cells would have existed and some still should. But there are none. Instead, there is a yawning gulf. On one side there are the tiny bacteria and archaea, collectively known as prokaryotes. On the other side there are the huge and unwieldy eukaryotes, the third great domain of life. A typical single-celled eukaryote, such as an amoeba, is about 15,000 times larger than a bacterium, with a genome to match.

Prokaryotes are little more than tiny bags of chemicals – complex bags, to be sure, but nothing compared with eukaryotic cells, which boast miniature organs called organelles, internal membranes, skeletons and transport systems. They are to prokaryotes what a human is to an amoeba.

And while bacteria never form anything more complex than chains or colonies of identical cells, eukaryotic cells aggregate and cooperate to make everything from seaweed to sequoias, aardvarks to zebras. All complex multicellular life forms – that is to say, pretty much every living thing you can see around you, and more besides – are eukaryotes.

All eukaryotes evolved from the same ancestor. Without that one-off event, life would still be stuck in its microbial rut. Bacteria and archaea cells just don't have what it takes to evolve into more complex forms.

So what happened? The critical event appears to have occurred about 2 billion years ago, when one simple cell somehow ended up inside another. The identity of the host cell isn't clear, but we know it engulfed a bacterium, which began to live and divide within it, like a squatter. The two somehow found a way to live together amicably, and eventually formed a symbiotic relationship called endosymbiosis.

Through co-evolution over countless generations, the endosymbionts eventually became an organelle called the mitochondrion. These stripped-down vestiges of their former bacterial selves evolved to have one key function: to supply the cell with energy. This was the critical step that allowed life to throw off its microbial shackles and evolve into endless forms most beautiful.

Turbo charge

Once they have mitochondria, cells can overcome a fundamental barrier that prevents bacteria and archaea from growing large. In a nutshell, there is a limit to how much energy microbes can produce. The cell's universal energy currency, ATP, is manufactured at the cell membrane. But as cells grow larger, their surface area to volume ratio drops and they have relatively less membrane to use. As they grow larger, their energy demands quickly overtake the supply. A cell with mitochondria (which have their own ATP-making membranes) can overcome this simply by adding more mitochondria – something that's easily done, as the mitochondria retain their bacterial ancestors' ability to clone themselves.

Awash with squadrons of mitochondria cranking out energy, early eukaryotes were free to grow larger and accumulate bigger and more complex genomes. And these expanded genomes provided the genetic raw material that permitted the evolution of ever more complex life.

Powered by the Sun

That was not the end of the story. Another round of endosymbiosis is thought to have created the chloroplast, the organelle that allows plants and algae to convert sunlight into sugar in the process called photosynthesis. The endosymbiont in this case was a photosynthetic bacterium, which first appeared on Earth about 2.8 billion years ago. The cell nucleus, where eukaryotes store the majority of their DNA, was another crucial invention. It may have been created by another endosymbiosis, possibly of a virus. Eukaryotic cells also acquired other organelles, such as the endoplasmic reticulum, where proteins are made, and the Golgi apparatus, which dispatches them to their destination, possibly by infolding of their cell membranes.

All of this set the scene for the emergence of complex, multicellular life forms. Admittedly, it took a while. The first large multicellular organisms were the Ediacarans, ocean-dwelling life forms which appeared about 700 million years ago and disappeared around the time of the so-called Cambrian Explosion 540 million years ago, when most of the familiar animal forms first evolved.

Nonetheless, the Ediacarans can trace their origins back to the evolution of mitochondria. And this seems to have hinged on a single fluke event – the acquisition of one simple cell by another. The bottom line is that while simple life appears to be a near inevitability, the evolution of complex life – including you and yours – is fantastically unlikely. That is the true miracle of life on Earth.

PLANET BORE

If life sometimes seems tedious, spare a thought for the inhabitants of Earth between 1.7 and 0.7 billion years ago. This unfathomably long stretch of time was so uneventful that biologists call it the 'boring billion'. The cause appears to have been geological rather than biological. The crust had solidified but plate tectonics had not really got going, leading to a long period of geological stasis devoid of rifting, volcanism, mountain building, continental drift and the other types of upheaval that often drive evolutionary change.

WHY DO WE HAVE SEX (APART FROM THE OBVIOUS REASON)?

THE BIRDS AND THE BEES AND, OF COURSE, THE FLEAS. Plants, fungi and amoebas too. It sometimes seems like sex is everywhere. But in biological terms, it is a minority pursuit. For the first 2 billion years of life on Earth, it didn't exist. Even now, the organisms that dominate the planet – bacteria and archaea – don't bother.

The origin of sex, then, is a bit of a mystery. And if its origins are hard to understand, its function is just as baffling.

At first sight that seems ludicrous. Surely sex has an obvious function: it generates variation, the raw material for evolution. The reshuffling and recombining of genetic information helps species adapt. It can also help spread beneficial genes throughout a population and eliminate harmful ones.

But there are big problems with this common-sense argument. The first is that sex is grossly inefficient. It makes much more sense to clone yourself. Cloning produces many more offspring than sex, which means that asexual species should rapidly drive sexual ones to extinction by dint of producing far more offspring competing for the same resources.

What's more, each clone has a combination of genes that has already been shown to be fit for purpose. Sex, by contrast, creates new, untested and possibly inferior combinations. In fact, sexual recombination disrupts favourable gene combinations more often than it generates them.

Sure, sex should be an advantage in the long term, over thousands and millions of years. Asexual species eventually accumulate mutations that they can't get rid of, and which drive them to extinction. But evolution doesn't work like that. It doesn't plan ahead. All it cares about is the here and now.

And the trials and tribulations don't end there. Sexual species have to find a mate, fight off rivals, and risk catching sexually transmitted diseases.

Finally, if sex is so beneficial, why is it that bacteria and archaea never evolved it, even though they do exchange bits of DNA from time to time? Conversely, if asexual reproduction is so great, why do almost all eukaryotes reproduce sexually at least some of the time? All this makes sex one of the biggest head-scratchers in biology.

For many years the best answer was the Red Queen hypothesis, a subtle variant on the 'sex means variety' explanation. This imagines an arms race between parasites and their hosts. The parasites' generation time is so short that they can out-evolve their hosts. By throwing up new mixtures of genes with each generation, sex enables at least a few individuals to survive. It is named after the Red Queen because, like Alice in *Through the Looking Glass*, we have to run fast just to stay in the same place.

Unfortunately, it does not solve the problem. Parasites give sex a decisive advantage only when parasite transmission is very high and their effects are very serious. Under normal circumstances, clones still win.

In recent years a new explanation has started to take hold. This is based on the discovery that all eukaryotes are, or at least were, sexual (there are plenty of species that multiply by cloning, but they evolved celibacy only very recently). The logical conclusion is that sex evolved very early on in the eukaryote lineage, in a common ancestor of all living eukaryotes around 2 billion years ago.

Aside from sex, the other thing that unites all eukaryotes is the possession of mitochondria, the cell's power supply. The new explanation claims that this is no coincidence: mitochondria made the evolution of sex inevitable. How so? The key point is that mitochondria have their own genomes. This is a remnant of the complete genome of the free-living bacterium that was engulfed at the dawn of eukaryote evolution. We know that as the two co-evolved, most of the genes were transferred to the host's genome. The symbiont also bombarded the host with parasitic jumping genes.

Love conquers all

In other words, the acquisition of mitochondria unleashed a bout of turbulent genetic disruption. Under such high mutation pressure, the balance was tilted and sex became more advantageous than asexual reproduction. Any early eukaryote that evolved it would have outcompeted its asexual rivals, which were succumbing to unsurvivable levels of mutation.

Mitochondria also explain why sex remains advantageous today. The mitochondrial genome encodes vital genes, but cannot do anything on its own. It relies on the nuclear genome to make proteins and replicate its DNA, for example. Close cooperation between the cell's two genomes is therefore vital to the functioning of the cell, especially in the crucial task of energy generation.

That cooperation is what sex ensures. Because the mitochondrial genome accumulates mutations at a higher rate than the nuclear genome – about 10 times faster in mammals – the accord between the two genomes gradually breaks down. We and our mitochondria are drifting apart, and though it is the mitochondria's fault, we are the ones who suffer. Sex resolves this disharmony by throwing out new combinations of nuclear genes that are more compatible with the mitochondria's needs.

That is the why of sex. The how, however, remains very unclear. The simplest eukaryotes – amoebas – have sex by splitting their genome in half and then cleaving themselves in two, with half a genome in each portion; these half-amoebas then merge with others to create new individuals. That may be how the first sex was done.

In broad brush terms, it is still how it is done. Sex just means ripping a genome in half and uniting it with another half-genome from someone else to create a new whole genome. Humans and most other animals achieve that by having two sexes, one of which dumps their half-genomes into the other through the act of copulation.

Who said romance was dead?

BIZARRE LOVE TRIANGLE

From a human perspective sexual reproduction is a one-on-one affair: there are males and there are females, and one of each is required to make a baby. The same is true of many other animals and plants, but this system is by no means universal. Some species of worms, sponges, molluscs and plants are hermaphrodites, meaning that any individual can mate with any other, or itself. And one species of ant has been found to have three sexes – a queen and two types of male. The queen has to mate with one type to make workers and the other to make queens. So the colony as a whole is the product of a threesome.

WHY ARE THERE SO MANY TYPES OF CREEPY CRAWLY?

IF YOU WANT TO LEAVE YOUR MARK ON THE WORLD BY DISCOVERING A NEW SPECIES OF ANIMAL, THERE'S REALLY ONLY ONE PLACE TO START: UNDER YOUR SETTEE, OR MAYBE ON A DUSTY WINDOW LEDGE. Look carefully and you may well discover a previously unknown species of insect.

Every year around 20,000 new species are described by scientists. The majority are invertebrates, and most of those – around 10,000 – are insects.

Love them or loathe them, insects are the success story of the animal kingdom. Three-quarters of all known animal species are insects, a staggering 1 million species with an estimated 4 to 5 million yet to be discovered. By contrast, there are fewer than 70,000 vertebrate species. There may be as many as 10 quintillion (10 billion billion) insects alive at any one time – that's more than a billion for each person on the planet. They were the first animals to conquer the land, have swarmed across every continent including Antarctica, and appear to be almost extinction-proof. In short, they are the most successful animals to ever walk or fly upon the Earth.

Petrified mini-forest

The oldest insect fossils ever found are 410 million years old, a time when life was making its first forays on to land.

The fossils come from a remarkable deposit found buried in a field near the village of Rhynie in Scotland. The Rhynie chert is a Lagerstätte, or fossil deposit of exceptional preservation. The fossils in it formed when hot, mineral-rich water flooded out of a volcanic spring and instantly petrified everything in its path.

The chert is teeming with fossilised life, mostly tiny plants. It also contains a zoo of early arthropods – animals with hard exoskeletons – including crustaceans, arachnids, mites and springtails. Insects were thought to be absent, but in 2004 palaeontologists spotted something under the microscope: perfectly preserved mouthparts that could only have come from an insect.

And not just any insect; the mouthparts were surprisingly modern-looking, which meant the insects were already quite advanced by the time the Rhynie chert formed. That pushed their origin way back.

As to what they evolved from, the initial suspects were myriapods – the group that includes millipedes and centipedes. Now, though, the smart money is on remipedes, blind aquatic crustaceans that today live exclusively in coastal caves. Similarities in their brains, nervous systems and many of their proteins all point to an ancient common ancestor. That suggests that insects evolved in the watery margins between sea and land.

Crawling on to land

The idea is backed up by a huge genetic study of insects and other arthropods, which firmly places the insects next to the crustaceans and pinpoints their origin to about 480 million years ago. That makes them among the first things ever to walk on land.

Colonising land was a formidable challenge, including dealing with dehydration, the effects of gravity, breathing air and daily extremes of temperature and sunlight. A tough exoskeleton would have helped, but it still took millions of years for truly terrestrial insects to evolve. Some of the most primitive species today, the jumping bristletails, still need moist soil to live in.

But the land offered big opportunities. There was plenty to eat and fewer predators than in the sea. Insect evolution really took off around 440 million years ago, with an explosion of species emerging.

Then came a development that would take them to another level: flight. The oldest fossilised insect wing is 324 million years old, but the mouthparts in the Rhynie chert are almost certainly from a flying insect, so we know that flight evolved very early.

For 200 million years, until pterosaurs appeared, insects ruled the skies. Wings gave them an enormous boost, helping them find food and mates, colonise new habitats, avoid predators and regulate their body temperature.

Insects had one more radical transformation in the pipeline – perhaps the most important of all. Among the fossilised remains of the swampy forests that covered the Earth around 300 million years ago are the earliest insects known to undergo complete metamorphosis – the process by which modern caterpillars turn into moths and butterflies, or maggots into blowflies.

Insects are fundamentally constrained by their inflexible exoskeleton. Until this point, they had grown via a series of stages, each followed by a moult, allowing miniature forms resembling the adult to get progressively bigger. Complete metamorphosis enabled the insects to divide their life cycle into distinct stages, with the larva dedicated to feeding and the adult to reproducing. This innovation was so successful that more than 8 in 10 insect species use it today, including hugely successful groups such as beetles, fleas, wasps, bees and ants.

Tough cookies

Metamorphosis is also what appears to make insects extinction-proof. Like everything else, insects were hit hard by the Permian mass extinction that killed off 90 per cent of all known species. Around half of insect families disappeared – but most of them were non-metamorphosing. The metamorphosers were hardly touched. The difference was probably the transitional phase between larva and adult, when insects retreat into a pupa. Pupae can tolerate all sorts of environmental assaults such as freezing and desiccation, making them very hardy during times of environmental stress.

When an asteroid impact put an end to the dinosaurs 65 million years ago, insects breezed through. Chances are, once we fly-by-night humans are gone, the fly-by-day-and-by-night insects will ride out whatever does us in and continue their reign as the most successful group of animals in the world.

HERE BE MONSTERS

Around 300 million years ago, insects suddenly ballooned in size. The predatory dragonfly-like *Meganeura*, for example, had a wingspan of up to 70 centimetres. The trigger was oxygen. Trees had recently evolved and, with no organisms to break down wood, were not rotting. As a result, oxygen levels reached 31 per cent, half as much again as today. Insects breathe through tiny tubes which carry oxygen to their tissues, which limits how big they can get. With more oxygen the limit is much larger.

Insects remained enormous until about 150 million years ago, when wingspans suddenly halved. The cause was probably the evolution of a new kind of flying insect-eater, birds.

WHEN DID THE AGE OF THE DINOSAURS BEGIN?

THE END, WHEN IT CAME, CAME SUDDENLY. An asteroid or comet 10 kilometres across slammed into the Gulf of Mexico, gouging a 180-kilometre crater and unleashing firestorms, eruptions and mega-tsunamis across the globe. The debris blocked out the Sun for years. The dinosaurs – and the other 75 per cent of life that went down with them – didn't stand a chance.

The story of the demise of the dinosaurs 65 million years ago is well known. But that of their origin is less so. Dinosaurs were the dominant animals on land for at least 135 million years, the longest reign of any group. Had the impact not happened, they might still be in control. Where did these magnificent beasts come from?

For years, palaeontologists thought that dinosaurs rose rapidly to prominence about 200 million years ago by virtue of being evolutionarily superior to their competitors. The Triassic period in which they first evolved was seen as little more than a dress rehearsal for the true age of dinosaurs – a kind of 'Jurassic-lite'.

We now know this isn't how it happened. The secret of the dinosaurs' success was luck: they were in the right place at the right time. And, like their demise, their origins and heyday were triggered by huge, catastrophic mass extinctions.

At the end of the Permian period 251 million years ago, more than 90 per cent of all life suddenly disappeared. The cause (or causes) of the wipeout is angrily debated, but there is no doubt about its devastating impact. Life itself nearly went extinct, leaving bleak and empty landscapes over the vast single continent of Pangaea. A few plants and large land animals somehow clung on, and over the next 50 million years they gradually refilled the empty planet with life.

The first to take advantage was a group of mammal-like reptiles called the synapsids. They dominated the Early Triassic, and gave rise to mammals. By the middle of the Triassic period a second group of reptilian Permian survivors called the diapsids were starting to take over. That's when things began to get monstrous.

Ruling reptiles

Some of these beasts took to the water and evolved into ichthyosaurs, plesiosaurs and the other familiar marine reptiles of kids' dinosaur books (though they were not dinosaurs). Another lot evolved into snakes and lizards. But the most interesting evolutionary action was taking place in a group of land animals called the archosaurs – the 'ruling reptiles'.

The classic view is that archosaurs evolved in the Middle Triassic and quickly gave rise to crocodiles, dinosaurs and the flying pterosaurs. They produced a few assorted 'others' too, but these were of no great significance. Almost as soon as dinosaurs evolved, they started throwing their weight around. Thanks to superior evolutionary adaptations, they quickly became the dominant land animals, making the Triassic the 'dawn of the dinosaurs'.

Or was it? It is true that the earliest dinosaurs are found in Middle Triassic rocks. The oldest come from a 230-million-year-old formation in the foothills of the Andes in Argentina.

The early birds

The first to be identified was *Herrerasaurus*, a very primitive two-legged meat-eater. Discovered in 1959, *Herrerasaurus* was found to belong to a group called the theropods, which ultimately gave rise to *T. rex*, *Velociraptor* and modern birds.

A few years later came *Eoraptor*, a member of the lineage that eventually evolved into the gigantic long-necked sauropod herbivores such as *Diplodocus* and *Apatosaurus*.

The discovery of *Pisanosaurus* completed the picture. It was a forerunner of the duck-billed dinosaurs, confirming that even at this early stage dinosaurs had split into their two major families: the 'lizard-hipped' saurischians, including theropods and sauropods, and the 'bird-hipped' ornithischians such as the duck-billed dinosaurs and the stegosaurs.

But more recent discoveries have challenged the idea that the dominance of the dinosaurs was already a done deal at this point. Far from being a supporting cast, the assorted 'others' were in fact the stars of the show, and dinosaurs hardly got a look-in until another extinction struck at the end of the Triassic. For whatever reason, this catastrophe hit the others hardest. All sorts of large, bizarre reptiles disappeared for ever. And much as the death of the dinosaurs cleared the way for the rise of mammals, so the Triassic reptiles' demise heralded the age of the dinosaurs. The Late Triassic was the heyday of the archosaurs.

The illusion of dinosaur dominance stemmed from the fact that fossils of Triassic land animals are rare and usually incomplete. When scientists found Triassic fossils that looked like they came from dinosaurs, they logically assumed that they were dinosaurs.

That included the rauisuchians, long-legged predators shaped like bears or lions. The largest stretched 7 metres. Some were bizarre, such as the sail-backed *Arizonasaurus*. Another dominant group of predators were the phytosaurs, long-bodied reptiles with narrow crocodilian jaws that looked a bit like modern gharials.

The most common plant-eaters were aetosaurs, low-slung animals up to 5 metres long with small heads and armoured bodies, built like the ankylosaurs of the dinosaur age.

For the next 10 million years the world belonged to these little-known animals, with dinosaurs playing bit parts. Then along came the Triassic–Jurassic mass extinction of 200 million years ago. It was one of the five most devastating extinctions of the past 500 million years but has attracted little attention, partly because there is no obvious trigger and partly because it claimed no charismatic victims.

Except that it did: the archosaurs. For some unknown reason they got absolutely hammered, leaving the dinosaurs to inherit the Earth.

OUT OF THE SHADOWS

The rise of the dinosaurs at the end of the Triassic period is similar to that of the mammals at the end of the Cretaceous period. After millions of years living in the shadows, they suddenly found they had the world largely to themselves, and they took advantage. Fossils show that while dinosaurs were few and far between in the Late Triassic, they dominated the Early Jurassic. The biggest dinosaur footprints jumped from 25 to 35 centimetres within just 30,000 years. This means that whatever made those tracks doubled in size over this time. This was the true dawn of the dinosaur age.

HOW DID EYES EVOLVE?

THEY APPEARED IN AN EVOLUTIONARY BLINK AND CHANGED THE RULES OF LIFE FOR EVER. Before eyes, life was gentler and tamer, dominated by sluggish soft-bodied creatures lolling around in the sea. The invention of the eye ushered in a more brutal and competitive world. Vision made it possible for animals to become active hunters, and sparked an evolutionary arms race that transformed the planet.

The first eyes appeared about 541 million years ago – at the very beginning of the Cambrian period when complex multicellular life really took off – in a group of now extinct animals called trilobites which looked a bit like large marine woodlice. Their eyes were compound, similar to those of modern insects. And their appearance in the fossil record is strikingly sudden. Trilobite ancestors from 544 million years ago don't have eyes.

So what happened in that magic million years? Surely eyes, with their interconnected assemblage of retina, lens, pupil and optic nerve, are just too complex to appear all of a sudden?

Designed by nature

The complexity of the eye has long been an evolutionary battleground. Ever since William Paley came up with the watchmaker analogy in 1802 – which claimed that something as complex as a watch must have a maker – creationists have used it to make the 'argument from design'. Eyes are so intricate, they say, that it stretches credibility to suggest they evolved through the selection and accumulation of random mutations.

Darwin was well aware of the argument. In *On the Origin of Species* he admitted that eyes were so complex that their evolution seemed 'absurd to the highest degree'. But he went on to argue convincingly that it only *seemed* absurd. Complex eyes could have evolved from very simple ones by natural selection as long as each gradation was useful. The key to the puzzle, Darwin said, was to find eyes of intermediate complexity in the animal kingdom that would demonstrate a possible path from simple to sophisticated.

Those intermediate forms have now been found. According to evolutionary biologists, it would have taken less than half a million years for the most rudimentary eye to evolve into a complex 'camera' eye like ours.

The first step is to evolve light-sensitive cells. This appears to be a trivial matter. Many single-celled organisms have eyespots made of light-sensitive pigments. Some can even swim towards or away from light. Such rudimentary light-sensing abilities confer an obvious survival advantage.

The next step was for multicellular organisms to concentrate their light-sensitive cells into a single location. Patches of photosensitive cells were probably common long before the Cambrian, allowing early animals to detect light and sense what direction it was coming from. Such rudimentary visual organs are still used by jellyfish and flatworms and other primitive groups, and are clearly better than nothing.

Out of the darkness

The simplest organisms with photosensitive patches are hydras – freshwater creatures related to jellyfish. They have no eyes but will contract into a ball when exposed to bright light. Hydras are interesting from an evolutionary perspective because their basic light-sensing equipment is very similar to that seen in other evolutionary lineages, including mammals. It is based on two types of protein: opsins, which change shape when light strikes them, and ion channels, which respond to the shape-shifting by generating an electrical signal. Genetic research suggests that all opsin/ion channel systems evolved from a common ancestor similar to hydras, pointing to a single evolutionary origin of all visual systems.

The next step is to evolve a small depression containing the light-sensitive cells. This makes it easier to discriminate the direction the light is coming from and hence sense movement. The deeper the pit, the sharper the discrimination.

Further improvement can then be made by narrowing the opening of the pit so that light enters through a small aperture, like a pinhole camera. With this sort of equipment it becomes possible for the retina to resolve images – a vast improvement on previous models. Pinhole camera eyes, lacking a lens and cornea, are found in the nautilus today.

The final big change is to evolve a lens. This probably started out as a protective layer of skin that grew over the opening. But it evolved into an optical instrument capable of focusing light on to the retina. Once that happened, the effectiveness of the eye as an imaging system went through the roof, from about 1 per cent to 100 per cent.

Eyes of this kind are still found in cubozoans, highly mobile and venomous marine predators similar to jellyfish. They have 24 eyes arranged in four clusters; 16 are simply light-sensitive pits, but one pair in each cluster is complex, with a sophisticated lens, retina, iris and cornea.

Hunt and destroy

Trilobites went down a slightly different route, evolving compound eyes with multiple lenses. But the basic sequence of events was the same.

Trilobites weren't the only animals to stumble across this invention, although they were the first. Biologists believe that eyes evolved independently on many, possibly hundreds, of occasions.

And what a difference they made. In the sightless world of the Early Cambrian, vision was tantamount to a superpower. Trilobites became the first active predators, able to seek out and chase down prey like no animal before. Unsurprisingly, their victims counter-evolved. Just a few million years later, eyes were everywhere and animals were more active and bristling with armour. This burst of evolutionary innovation is what we now know as the Cambrian Explosion.

However, sight is not universal. Of 37 phyla of multicellular animals, only six have evolved it. But these six – including our own phylum, chordates, plus arthropods and molluscs – are the most abundant, widespread and successful animals on the planet.

THE EYE OF THE PLANKTON

Eyes are remarkable organs, but possibly the most remarkable of all is possessed by a single-celled animal called *Erythropsidinium*. Around a third of its tiny body is occupied by a structure called an ocelloid, which, despite being microscopic, is staggeringly similar to the sophisticated camera-like eyes of vertebrates. At the front is a clear sphere rather like a cornea. At the back is a dark, hemispherical structure where the light is detected. *Erythropsidinium* apparently uses its eye to locate prey despite having no nervous system. Exactly what it 'sees' is anyone's guess.

WHY DO WE SLEEP?

WITHIN A FEW HOURS OF READING THIS YOU WILL LOSE CONSCIOUSNESS AND ENTER THE TWILIGHT ZONE. For the next few hours your brain will cycle between two wildly different states, deep sleep and rapid eye movement sleep, or REM. For much of that time you will not be totally unconscious but engaged in the bizarre nocturnal state called dreaming.

We spend about a third of our lives asleep, and it is clearly very important. If deprived of sleep for too long, we fall ill; rats kept awake 24/7 die within three weeks. Yet despite more than 60 years of intense study, we still don't really know what it is for.

It isn't for lack of trying. Sleep scientists have come up with dozens of hypotheses about its function. These range from keeping us out of harm's way to saving energy, repairing our bodies and brains, tuning our immune system, processing information, regulating emotions and consolidating memory. Each has strengths – and weaknesses, too. Most sleep researchers accept that sleep has many functions and that all of these hypotheses may be true to some extent.

The lack of an accepted explanation, however, is not just

frustrating for sleep researchers. It has also made the evolutionary origin of sleep very hard to discern. It must be very old: all animals with a complex nervous system do it, including mammals, birds, reptiles and fish. We know that dinosaurs slept: in 2004 palaeontologists in China discovered the bones of a 125-million-year-old dinosaur with its head tucked under its forelimb, exactly like a sleeping bird with its head under a wing. It is also possible to identify sleep-like states in animals without a complex nervous system, including insects, scorpions, worms and some crustaceans.

Sleep may even be an inherent property of nerve cells. Groups of neurons grown in a Petri dish spontaneously enter a state very like sleep. Deprive them of their shut-eye and they go haywire, firing rapidly and randomly in an epileptic-like frenzy.

Even microorganisms, which lack a nervous system altogether, have daily cycles of activity and inactivity driven by internal body clocks. The origins of sleep might therefore date back to the dawn of life around 4 billion years ago.

Another stumbling block is that sleep is not just one thing but two. The first is called deep or slow wave sleep because it is characterised by long, lazy waves of undulating electrical activity synchronised across the whole brain. The second is rapid eye movement (REM) sleep, which could hardly be more different. It is characterised by frantic brain activity that looks very much like wakefulness. It also has obvious physical signs: the rapid flickering motion of the eyeballs and near total muscle paralysis thought to prevent you acting out your dreams.

REM sleep is found only in mammals and birds. They last shared a common ancestor about 300 million years ago, which could mean that REM sleep evolved before then. However, that common ancestor also gave rise to REM-less reptiles, which suggests that birds and mammals evolved REM sleep independently.

Where your mind goes at night

REM sleep is also when we do most of our dreaming, the function and origins of which sleep scientists have made much more progress in understanding.

Sigmund Freud was the first to suggest that the content of dreams can be influenced by waking experiences. He called these 'day residue'. Freud's ideas about dreaming have fallen largely out of favour, but this one – now known as the continuity hypothesis – is still influential.

Dreams seem to hold a mirror to our waking lives. They often reflect recent experiences, particularly new ones. Someone who has just played Tetris for the first time may dream of oblong shapes falling from the sky, for example.

The link between waking and dreaming has also been observed directly by brain scanners, which reveal the dreaming brain replaying patterns of activity that were seen during earlier waking experiences.

Experiences seem to enter our dreams in two separate stages. They first reappear the night after the event itself, and then again between five and seven days later, thus supporting the idea that one of the functions of sleep is to process memories and integrate them into long-term storage.

We do not simply replay events in our dreams, however. They

are fragmented, combined with older memories and jumbled into bizarre and emotionally charged narratives featuring impossible events, places and characters. This may simply be due to the brain activity required for memory processing. Visual areas are very active, as are emotional centres in the amygdala, thalamus and the brainstem. At the same time, the areas which deal with rational thought and attention are quiet.

But memory processing cannot be the whole story. Dream reports gathered from people born with disabilities contain elements that they have never actually experienced. Many deaf people have dreams in which they can hear and understand spoken language; those who cannot speak in real life find their voices. People born paralysed often walk, run or swim. This suggests that, for some reason, the brain is genetically programmed to generate experiences that we can expect to encounter during our lives.

Something similar may explain nightmares. About two-thirds of dreams involve a threat, often a fraught encounter such as running away from an assailant or getting into a fight. Such encounters are even more common among children, and often feature dangerous animals. One explanation for this is that the brain conjures up dreams to simulate challenges that we may meet in real life – or that our distant ancestors would have experienced – allowing us to practise dealing with them.

So when you lose consciousness tonight, beware. The twilight world is full of mystery and danger.

NOCTURNAL JOURNEYS

REM sleep is often called 'dreaming sleep' and it is when most dreams occur. But we also dream during other stages of sleep. When researchers monitor the brains of sleeping people, they can see that dreaming also occurs in non-REM sleep, though these dreams are shorter, less vivid and less complex than REM dreams.

Another type of dream occurs at the boundary of sleep and wakefulness. These fleeting hypnagogic dreams have a hallucinatory quality, and can sometimes be the doorway to yet another type, lucid dreaming. This is a thrilling and sought-after state of consciousness in which you become aware that you are dreaming and can exert some control over what happens. Talk about living the dream.

HOW DID APES BECOME HUMAN?

MANY PARENTS DREAD THE MOMENT WHEN A CHILD ASKS WHERE THEY CAME FROM. Darwin found the subject awkward too: *On the Origin of Species* makes almost no mention of human evolution.

Darwin was being tactful. The idea of evolution in any form was controversial enough in the middle of the nineteenth century. Claiming that humanity had been shaped by evolution was explosive, as Darwin found when he published a book all about it in 1871.

There was also a scientific barrier. Darwin had access to almost no fossil evidence that might indicate how, when or even where humans evolved.

In the intervening years the human – or hominin, to use the proper term – fossil record has expanded enormously. There is still much to discover, but the broad picture of our evolution is largely in place. We know that our evolutionary tree first sprouted in Africa. We are sure that our closest living relatives are chimpanzees, and that our lineage split from theirs about 7 million years ago.

The road to humanity was a long one, however. Nearly 4 million years later, our ancestors were still very ape-like. Lucy, a famous 3.2-million-year-old human ancestor discovered in Ethiopia, had a small, chimp-sized brain and long arms that suggest her species still spent a lot of time up trees, perhaps retreating to the branches at night as chimps still do. But she did have one defining human trait: she walked on two legs.

Lucy belongs to a group called the australopiths. In the 40 years since her partial skeleton was discovered, fragmentary remains of even older fossils have been found, some dating back 7 million years. These follow the same pattern: they had chimp-like features and tiny brains but probably walked on two legs.

We also know that australopiths probably made simple stone tools. These advances aside, australopiths weren't that different from other apes.

Only with the appearance of true humans – the genus *Homo* – did hominins begin to look and behave a little more like we do. Few now doubt that our genus evolved from a species of australopith, although exactly which one is a matter of debate. It was probably Lucy's species *Australopithecus afarensis*, but a South African species, *Australopithecus sediba*, is also a candidate. It doesn't help that this transition probably occurred between 2 and 3 million years ago, a time interval with a very poor hominin fossil record.

The earliest species of *Homo* are known from only a few bone fragments, which makes them difficult to study. Some doubt that they belong in our genus, preferring to label them as australopiths. The first well-established *Homo*, and the first that we would recognise as looking a bit like us, appeared about 1.9 million years ago. It is named *Homo erectus*.

Erectus was unlike earlier hominins. It had come down from the trees completely and also shared our wanderlust: all earlier hominins are known only from Africa, but *Homo erectus* fossils have been discovered in Europe and Asia too.

The toolmaker

Homo erectus was also an innovator. It produced far more sophisticated tools than had any of its predecessors, and was probably the first to control fire. Some researchers think that it invented cooking, improving the quality of its diet and leading to an energy surplus that allowed bigger brains to evolve. It is certainly true that the brain size of *Homo erectus* grew dramatically during the species' 1.5-million-year existence. Some of the very earliest individuals had a brain volume below 600 cubic centimetres, not much larger than an australopith, but some later individuals had brains with a volume of 900 cubic centimetres.

Successful though *Homo erectus* was, it still lacked some key human traits: for instance, its anatomy suggests it was probably incapable of speech.

The next hominin to appear was *Homo heidelbergensis*. It evolved from a *Homo erectus* population in Africa about 600,000 years ago. This species' hyoid – a small bone with an important role in our vocal apparatus – is virtually indistinguishable from ours, and its ear anatomy suggests it would have been sensitive to speech.

According to some interpretations, *Homo heidelbergensis* gave rise to our species, *Homo sapiens*, about 200,000 years ago in

Africa. Separate populations of *Homo heidelbergensis* living in Eurasia evolved too, becoming the Neanderthals in the west and a still enigmatic group called the Denisovans in the east.

Last man standing

Within the last 100,000 years or so, the most recent chapter in our story unfolded. Modern humans spread throughout the world and Neanderthals and Denisovans disappeared. Exactly why they went extinct is another great mystery, but it seems likely that our species played its part. Interactions weren't entirely hostile, though: DNA evidence shows that modern humans sometimes interbred with both Neanderthals and Denisovans.

There is still much we do not know, and new fossils have the potential to change the story. Three new extinct hominins have been discovered in the past decade or so, including *Australopithecus sediba* and the enigmatic and not yet well-dated *Homo naledi*, also in South Africa. Strangest of all is the tiny 'hobbit' *Homo floresiensis*, which lived in Indonesia until about 12,000 years ago and appears to have been a separate species.

For 7 million years our lineage had shared the planet with at least one other species of hominid. With the hobbit gone, *Homo sapiens* stood alone.

TWO LEGS GOOD

Bipedalism is one of the defining traits of our species. It's still a mystery exactly when our ancestors began to walk on two legs, but it brought advantages that helped to propel us across the world and eventually beyond. Travelling long distances is more efficient on two legs and an upright stance affords a better view of predators and reduces overall exposure to the midday sun. Perhaps most important of all it freed up our hands, allowing them to evolve into the multi-purpose, opposable-thumbed tools that were also crucial to our evolutionary success.

WHAT WERE THE FIRST WORDS?

IF YOU MET ANOTHER HUMAN BEING AT RANDOM, CHANCES ARE YOU WOULD BE UNABLE TO COMMUNICATE WITH THEM BEYOND A FEW GRUNTS AND GESTURES. At the last count there were almost 7,000 spoken languages in the world; the most common, Mandarin Chinese, is spoken by just 14 per cent of people. The rarest can count their speakers on the fingers of one hand.

Despite this diversity, there is a commonality among languages. All cultures have them, and linguists believe that deep down they are all essentially the same. The human brain is born language-ready, with an in-built program that is able to learn whatever mother tongue it happens to be born into.

The origin of this unique skill was clearly a major event, but is extremely difficult to pin down. Words do not fossilise, and the oldest written language is a mere 6,000 years old. But that doesn't mean that the origin of language is a closed book.

Speak your mind

Linguists define language as any system which allows thoughts to be freely expressed as signals and the signals to be converted back into thoughts. This sets human language apart from all other animal communication systems. Though many animals possess some of the elements of language, only we have the full set: the ability to use signals and learn new ones; the ability to articulate signals as words; and the ability to string words together, according to the rules of syntax and grammar, to convey ideas about anything under the Sun.

Most theorists agree that early humans did not acquire this full package all at once, but passed through multiple stages en route to modern language. For much of human prehistory, our ancestors possessed some of, but not all, the components of language. Such a system is termed a protolanguage.

One obvious possibility is that protolanguage was made up of words. This 'lexical protolanguage' model suggests that early humans used words but did not arrange them into sentences. It parallels language development in children, who start out with single-word utterances, move on to a two-word stage, and then begin forming more complex sentences.

But in that case, where did the first words come from? Words are only useful if they have a shared meaning, as two people who speak different languages soon find.

Another hypothesis focuses on the origins of vocal learning – the capacity to produce complex strings of sound. Many animals including whales and songbirds can do this, but their vocalisations do not communicate detailed information. Instead they are displays of virtuosity designed to attract a mate or claim territory. Based on this, linguists have suggested that protolanguage resembled whale or bird song and evolved for the purposes of sexual selection or territoriality. Only later did the notes and syllables take on meaning. One virtue of this is that it also explains the origin of music, another universal characteristic of our species.

A third possibility is that language started as gestures. Evidence for this comes from apes, which use manual gestures to convey information and can be taught human sign languages to a fairly high level. However, gestural models face the difficulty of explaining why we switched to speech. It may have been due to the need to communicate in darkness, or because hands became occupied by tools.

What about when it happened? Again, it is impossible to say for sure, but we can make educated guesses. We are pretty sure that our closest relatives, the Neanderthals, had full-blown language. They had the same neural connections to the tongue, diaphragm and chest that allow us to control our breathing and articulate intricate sounds. They also shared our version of a gene called *FOXP2*, which is crucial for forming the complex motor memories required for speech. Assuming this gene variant arose just once, speech must pre-date the divergence of the *Homo sapiens* and Neanderthal lines around 500,000 years ago.

For older ancestors, the fossil record does not speak so eloquently. It is plausible that our lineage had the gift of the gab 600,000 years ago, when *Homo heidelbergensis* appeared in Europe. Fossilised remains show they had lost a balloon-like organ connected to the voice box that allows other primates to produce booming noises to intimidate their opponents. This would have removed a major barrier to producing speech.

Spoken like a true human

Language could be even more ancient. You have to go back 1.6 million years to find an ancestor that lacks human-like neural connections, suggesting that even very early humans were capable of speech. But the protolanguage hypotheses muddy the waters a bit. If language started with gestures, then hominins could have been using sign language even earlier. Conversely, if it started as music, then 'speech' adaptations could have been for producing something like whale song, with little specific information.

Even so, *Homo heidelbergensis* and Neanderthals made complex tools and hunted dangerous animals – activities that would have been very difficult to coordinate without at least some language.

The same can probably be said for *Homo erectus*, which had a brain not much smaller than ours, suggesting a capacity for intelligence and culture. Their stone tools were vastly more sophisticated than anything that preceded them. But tellingly, their tools reached a kind of stasis – the all-purpose hand axes persisted unchanged for a million years. This suggests that they did not have full language, which would have accelerated cultural and technological change. Hence they might have had some, but not all, of the linguistic capacities of modern humans – in other words, a protolanguage.

TALK TO THE ANIMALS

Many animals have communication systems that almost but not quite resemble language. Vervet monkeys, for example, have different alarm calls for different predators including 'eagle' (which sees them running for cover) and 'leopard' (which propels them up the nearest tree). However, vervets do not invent new alarm calls and so their system cannot be considered a language.

Similarly, many species produce complex strings of sounds. But this is not the same as possessing language. Talking parrots do not understand what they say, nor what we say back to them. And while the vocalisations of birds or whales can rival human speech in complexity, they usually convey only a very simple message: 'I'm over here, I can sing really well, and I'm looking for a mate.'

WHY DO WE MAKE FRIENDS?

HOW BIG IS YOURS? If you're a typical human, then it is probably about 1,500.

That's the size of your social network, by the way – the sum total of people who are not strangers to you. That includes family, colleagues, acquaintances, people you recognise but don't really know – and, of course, friends. Of those 1,500 non-strangers, about 50 count as true friends, including 10 close friends and 5 intimate ones.

Of all our social relationships, friendship is the most peculiar. You can't choose your family or colleagues, and acquaintances are of little consequence. But friends are different. Most animals only cooperate with their close relatives, so why do we spend so much time and effort cultivating close relationships with people who are not related to us? And how do we choose who to befriend?

Friendship may seem peculiarly human but it actually has deep evolutionary roots. Many other species of mammal also build close personal ties with unrelated individuals, including the great apes and many other primates, elephants, horses, whales, camels and dolphins.

What they all have in common is that they live in large social groups with complex hierarchies. Living in groups like this has major benefits but also creates tensions, and that is where friends come in. To put not too fine a point on it, friends are a personal protection squad who come to our aid when we need help. Overlapping circles of friends within a large group create coalitions and alliances that help to maintain overall group stability.

For humans and other animals who live in large social groups, friends are not a nice-to-have, they are a biological necessity. A good social life helps to keep us physically and mentally healthy, while social isolation is stressful and makes us susceptible to illness.

Unsurprisingly, then, we have a powerful biological urge to make and keep friends. Like food and sex, this drive is controlled by the brain's reward centres, which give us shots of various feel-good chemicals in response to sociable behaviour.

One is oxytocin, the so-called 'cuddle chemical' which is a key reinforcer of the mother–infant bond. It is also released in response to positive social contact with other people. The resulting warm feeling is a reward that encourages you to see that person again.

Another is a group of neurotransmitters called endorphins. These are released in response to mild stressors such as exercise and act to damp pain and create a general feeling of wellbeing. They are also released in response to social contact, especially cooperation.

If you put somebody in a boat and ask them to row it from A to B, their brain releases endorphins in response to the physical effort. But if you put two people in the boat and ask them to row it together, their brains release more endorphins even though they exert less physical effort.

I scratch your back ...

Many species build and maintain friendships by grooming one another. Baboons, for example, spend hours every day picking parasites and dirt out of each other's fur, which triggers mutual release of those feel-good and trust-building chemicals oxytocin and endorphins.

But grooming takes time, which limits the number of social relations an individual can maintain. In monkeys and apes, the upper limit is about 50. This limit is also restricted by brain size. Navigating a sea of overlapping, shifting relations takes brain power, especially the ability to understand others' states of mind.

Apes can do this to some extent, being able to hold thoughts such as 'I know that she is friends with her', so-called 'third-order intentionality'. But humans are much smarter, and are able to juggle fifth- or even sixth-order intentionality: 'I know that you think that he wonders whether she is afraid that he has got it in for me.'

This mind-reading power has allowed us to transcend the upper limit and maintain social circles of about 150 – known as 'Dunbar's number' after the Oxford evolutionary biologist Robin Dunbar who calculated it. Our intelligence has also allowed us to evolve proxies for actually combing through body hair and picking out nits, which allow us to 'groom' more than one friend at a time. They include laughter, singing and jokes and, crucially, gossip.

Smart socialisers

This link between brain size and group size – sometimes referred to as the social brain hypothesis – also applies to individuals. Macaques and humans with larger brains also tend to have more friends. The absolute upper limit for an individual human appears to be about 250.

The exact membership of your social circle is largely down to chance: where you live, where you went to school, where you work. But from within that 150 or so, how do we end up becoming special friends with a select few?

On the surface the answer is quite simple. We form close friendships with people who are like us, with similar personalities, interests, beliefs, tastes, sense of humour and so on. But this simplicity belies a deeper connection. People turn out to be more genetically related to their close friends than to random strangers. A typical close friend is about as close as a fourth cousin; that is, someone you share a great-great-great-grandparent with.

No one knows how we recognise genetically similar people to befriend. It could be similarities in appearance, voice, smell or personality. But the fact that our friends are also distant relations helps to answer the million-dollar question of why we invest so much time in them. Evolution ought to prioritise cooperation with kin, as this helps us to fulfil life's prime directive, which is to pass on our own genes to the next generation, even if only by proxy. But if our buddies are distant kin, then it turns out that is what friends are really for.

ABSENT FRIENDS

Making friends is one thing; keeping them is another. Friendships are highly susceptible to decay if not maintained. Failure to see a friend for a year causes the quality of that friendship to decline by about a third, with a resultant drop down the hierarchy of closeness. Family relationships, by contrast, are much more resilient. As a result, the family part of our social network remains essentially constant throughout life whereas the friendship component undergoes considerable churn, with about 20 per cent turnover every few years.

WHERE DOES BELLY-BUTTON FLUFF COME FROM?

As an exercise in scientific navel-gazing, Georg Steinhauser's experiment takes some beating. Starting in 2005, Steinhauser – then a chemist at the Vienna University of Technology – collected pieces of belly-button fluff from his navel and recorded their colour and weight. Over the next three years he collected 503 pieces of lint, weighing almost a gram in total. At one point he shaved his navel. He asked male friends, colleagues and family members about their own lint production, and the general state of their navels.

He sent some of his lint off for chemical analysis, and eventually published his findings in a scientific journal. And all in the interests of answering the question: Why do some people find so much fuzz in their belly buttons?

Fuzzy ideas

Steinhauser had noticed his own prolific productivity in this department in his early twenties. He searched the scientific literature but only came across one enigmatic piece in the journal *Nature*. Under the title 'Another Matter: Navel Fluff', it showed three black and white pictures of what looked like cotton wool labelled 'sailor (while at sea)', 'farmer' and 'architect', with no further explanation. Two weeks later it published a correction: same pictures, same labels, but the fluff from sailor and architect swapped.

Steinhauser was eventually galvanised and took the lint in his own hands after reading a book called *Why Do Men Have Nipples?* which said that the fluff question – i.e. why do some belly buttons collect lint and others don't – could not be answered.

Take a quick look in your own belly button. Any fluff? If so, what colour is it? And how would you describe your navel: hairy or smooth?

Steinhauser's first observation was that, as a rule, his fluff was the same colour as the top he was wearing that day, which led him to suspect that it was derived from his clothing. Chemical analysis pointed in the same direction. On a day when he had worn a plain white cotton T-shirt he found that the accumulation was largely cellulose – the protein that makes up cotton – plus a smattering of extra nitrogen and sulphur compounds. These contaminants were probably dead skin, dust, fat, proteins and sweat, he concluded.

Next he probed the role of navel hair, something he was abundantly blessed with. From his survey of the contents of other men's belly buttons, he concluded that 'the existence of abdominal hair was a major prerequisite for the accumulation of navel fluff'. Shaving off his abdominal hair stopped the fluff from accumulating until the hair grew back. He also noted that small pieces of fuzz first appeared among the hair and then ended up in the navel at the end of the day.

Steinhauser had his grand unified theory of belly-button fluff. Hair is scaly and so abrades fibres from the fabric of the garment. The scales also act like barbed hooks, dragging the fibres towards the belly button. Abdominal hair often seems to grow in concentric circles around the navel, which enhances the belly button-ward motion, like matter spiralling into a black hole. Once over the event horizon, the fibres are 'compacted to a felt-like matter'. A T-shirt worn 100 times would lose approximately 0.1 per cent of its mass to navel fluff, Steinhauser calculated.

Abdominal pains

The research was just a bit of light relief from Steinhauser's real research on the chemistry and physics of radioactive elements. But sometimes belly-button deposits are no laughing matter. Soon after Steinhauser published his research, doctors in Nebraska reported a case of a 55-year-old obese woman with a rare condition called omphalitis, or inflammation of the umbilical cavity. She had been bleeding from it for four months. When the doctors examined her they saw a 'dark, rounded mass' which they suspected was a tumour. But it turned out to be a ball of lint nearly a centimetre across. They pulled it out, and she was cured.

Some of the human body's other gubbins have similarly obvious but nonetheless interesting origins. Earwax, or cerumen as doctors call it, is mostly dead skin plus an oily secretion called sebum and a watery one from sweat glands. People fall into two distinct earwax camps: wet and dry. Wet earwax is orange-brown and sticky; the dry type is translucent and scaly, like dead skin. That is because it *is* dead skin. People with dry earwax don't make the oily stuff and their 'wax' is just keratin and dirt.

Getting up your nose

A few years ago geneticists discovered that dry earwax is caused by a recessive mutation in a single gene, *ABCC11*. Earwax consistency therefore joins a number of other either/or human traits that are controlled by a single gene, such as tongue rolling, attached/detached earlobes, and ability/inability to smell freesias.

Bogeys, meanwhile, are the crusty remnants of the nasal mucus that is constantly being produced to protect the lining of the nasal cavity. Most is swept down into the pharynx by cilia and swallowed (so you are a snot-eater whether you like it or not), but some gets stuck in the nostrils and dries out. Bogeys are green because the mucus contains an antimicrobial enzyme called myeloperoxidase, which is made green by an iron-containing heme group. In an informal survey, almost half of adults admitted to eating their bogeys and enjoying them.

It's a similar story for the sleep – or rheum – that you find in and around your eyes when you wake up. This is basically an eye bogey – dried yellowish mucus that accumulates in the corners of your eyes when they are closed.

Basically, wherever your body has a cosy little corner where nothing much happens, something rather revolting is likely to accumulate.

DO YOU EAT YOUR BOGEYS?

Go on, you can admit it. In an informal survey carried out by food writer Stefan Gates, 44 per cent of adults confessed to eating their dried snot and enjoying it. Among the pantheon of weird human behaviours, this is right up there. Why do it? Nobody eats their earwax, sleep or belly-button fluff. One idea is that eating snot may be beneficial to the immune system. An Austrian doctor called Friedrich Bischinger reportedly tells parents to encourage their children to do it.

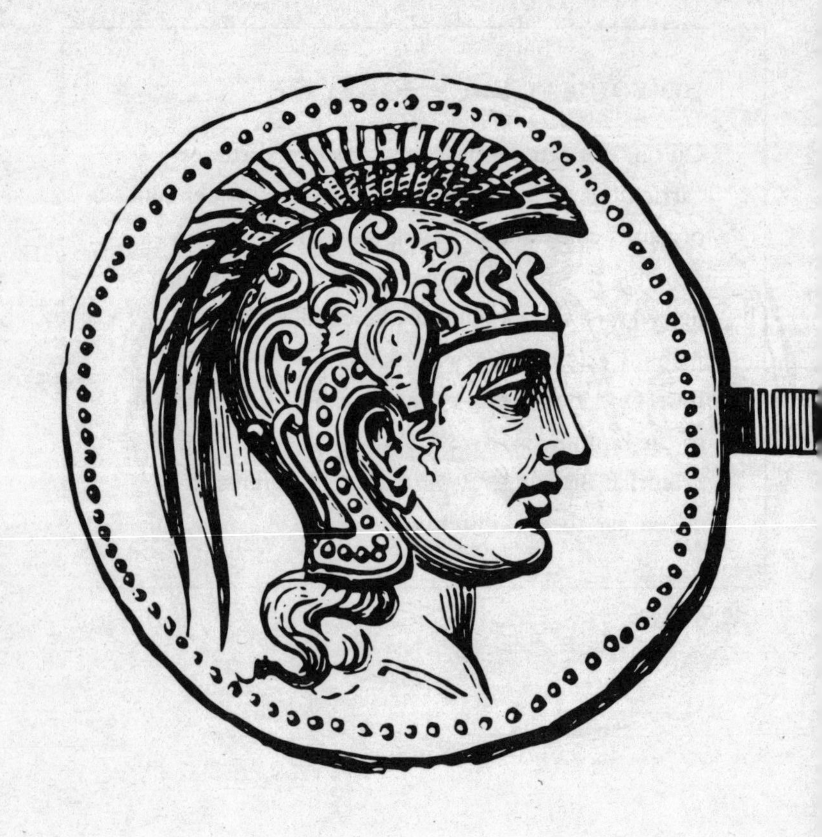

PART FOUR

CIVILISATION

WHEN DID WE START LIVING IN CITIES?

IN 2014 HUMANITY OFFICIALLY BECAME AN URBAN SPECIES. For the first time ever more people lived in cities and towns than in rural areas – an incredibly rapid transition, given that in 1960 only a third of people lived in urban areas.

Until about 5,500 years ago, nobody lived in cities. Villages had existed for millennia, but none had made the transition in size or, crucially, complexity to attain full-blown city status. How and why did that quintessential act of civilisation happen?

The dawn of urbanism

To answer that, you need to cross the threshold of recorded history and go back more than 6,000 years, to a time before writing was invented and when stone tools had yet to be fully displaced by metal.

According to the orthodox view, the world's first cities originated in Mesopotamia, a tract of fertile land between the Euphrates and Tigris rivers. The daddy of them all was the Sumerian city of Uruk. Situated on the banks of the Euphrates, it probably started life as one of the first permanent settlements that started springing up in the Fertile Crescent in about 8000 BC. By 3500 BC it had grown into a true city, covering around 2.5 square kilometres and housing about 50,000 people, most of whom would have been strangers to one another.

If you could travel back in time to Uruk you would recognise it as a city. Among the most obvious urban features would have been large buildings, which, unsurprisingly, are one of the defining features of cities. Not just any old large buildings, though, but non-religious public buildings, which are indicative of systems of government and accounting.

You would also see evidence of functional 'zoning' – clear differentiation between administrative centres, residential quarters, markets, rubbish dumps and so on. Fortification, too, would have been prominent, indicating wealth that was worth defending. Earlier large settlements, such as seventh millennium BC Çatalhöyük in modern-day Turkey, fail these tests of cityhood.

One thing that would have been conspicuously absent at Uruk, however, was a sophisticated transport network. Asses had been domesticated but the wheel was still a distant prospect.

Uruk was the end product of a long process that started when early farmers first settled in villages to be near their crops and animals. These settlements could grow very large, as at Çatalhöyük. Eventually, agricultural surpluses allowed some people to give up farming and take up other occupations such as metalwork. This division of labour led gradually but inexorably to the formation of genuine urban zoning, as specialist craftspeople congregated near others who possessed skills that they lacked.

Uruk isn't just considered the first true city. It is also seen as the origin of the wave of urbanisation that swept across Mesopotamia between about 3400 and 3100 BC, as people from the south colonised the entire region and built cities in the image of Uruk.

Northern powerhouses

But in the past couple of decades, the primacy of Uruk has been challenged by discoveries in the supposedly backward northern provinces. At least two sites have yielded clear evidence of urbanisation before the oldest evidence from Uruk. These findings have prompted some archaeologists to propose a serious rethink.

One site is Tell Hamoukar in eastern Syria. Western archaeologists have known about the location since the 1920s but marked it as a 'secondary' city, a product of the wave of urbanisation spreading north from Uruk. However, more recent excavations suggest that the site had already attained an advanced state of urbanisation before Uruk started throwing its weight around.

Even as early as 3700 BC, Hamoukar covered about 12 hectares and was enclosed by a defensive wall. Inside the wall are the remains of a large, secular and possibly public building – maybe a canteen of some kind. Archaeologists also found numerous kinds of 'seal stamps', which were used to impress a symbol on wet clay or bitumen to keep track of goods. Seal stamps are well known from Uruk and are widely accepted as solid evidence of urbanism because they indicate systems of accounting.

Way back in 3700 BC, then, Hamoukar already had many of the telltale features of early urban life. Yet there is no sign of a southern influence: Uruk-style pottery only starts showing up around 3200 BC. It seems as though the people of Tell Hamoukar were living in a city, independent of Uruk and possibly even unaware of its existence, at least 500 years before cities supposedly arrived in their area.

Lost in time

An even older possible city is Tell Brak, also in Syria. It too has the remains of a spectacularly large, spectacularly old building, with walls a metre and a half thick and a huge doorway opening out on to a courtyard. When archaeologists dated it they found it was more than 6,000 years old. Seal stamps also turned up, and much of the pottery from the time bears evidence of some kind of accounting and administration system, which suggests that the people of Brak were using complex administrative technology well before Uruk.

Most archaeologists agree that the case is not yet closed, however. The comparable period in the south is not well known. It is possible that southern Mesopotamia had fully fledged cities before 4000 BC, but no one has yet dug deep enough to find them. The whole region has been off-limits to archaeology for years due to war and unrest. Work probably won't resume for several years, if ever. In the meantime, archaeologists are hoping that the site has not been looted beyond repair, wiping the evidence of the origins of civilisation off the map for ever.

STONE AGE SUBURBIA

When it was discovered in 1958, Çatalhöyük captured the world's imagination. The 13-hectare site contained hundreds of buildings packed tightly together and was once home to an estimated 10,000 people. It looked like a city, but was incredibly old: the earliest remains were about 9,000 years old. However, it lacked a key feature found in true cities: functional differentiation between zones. Çatalhöyük was just lots and lots of houses and rubbish dumps, like an ancient suburbia. The people, it seems, carried out all their activities from home, even burying their dead under the floor. True cities did not emerge for another 1,500 years.

WHY DO WE TREAT WORTHLESS BITS OF PAPER LIKE GOLD?

EMPTY OUT THE CONTENTS OF YOUR WALLET. Coins, notes, cards: they may represent your wealth (or lack of it) and allow you to buy things, but they are really just figments of your imagination. Money has no intrinsic value. Instead, like Tinkerbell – the fairy in *Peter Pan* who dies if children's belief falters – money lives and dies on our collective faith in it.

To understand what underpins that belief, it helps to understand how and why money was invented.

It started with barter. Historians are divided on exactly where and when, but barter seems to have been in full swing in Mesopotamia by 8000 BC.

Barter was more convenient than fighting over resources, but it had its flaws. First, it depended on a surplus. Worse, it hinged on 'the coincidence of wants': say you have two surplus sheep and need a cow, you need to find someone with a spare cow and a need for two sheep. The solution was to build up a chain: barter your sheep for some grain and use that to pay for the cow.

Chains like these identified the items that had the most

universal value, and because these could be traded for almost anything, they became early 'commodity' money. Salt was one; useful materials such as metals were another. Because everyone could agree on their value, these commodities came to be the benchmark for the value of everything, and gradually came to be used principally as a store of value and a medium of exchange. In other words, money.

Coin of the realm

The obvious next step was to create standardised units of money. The first to do so were the Chinese, who minted metal coins around 1000 BC. Coins were portable and durable and they caught on, but they also had a serious design flaw. Early coins were made of precious metals which made them vulnerable to clipping – shaving off slivers and then passing off the coins at their face value. That is why gold's value was eventually transferred to paper.

In the thirteenth century AD Chinese merchants began depositing their surplus coins with other merchants in exchange for receipts stamped with a promise that this piece of paper corresponded to a specific number of gold coins. They could redeem these 'promissory notes' whenever they wanted. The Chinese state soon started issuing its own notes, and by 1274 China had a nationwide paper currency.

This path was accidentally recapitulated in seventeenth-century England by goldsmiths who stored gold in their vaults. They issued receipts certifying amount and purity and promising to pay back exactly what had been deposited. These receipts inevitably took on a life of their own. They were used to settle debts and make purchases and started circulating as a form of paper currency.

Neither a borrower ...

It didn't take long for the goldsmiths to cotton on that they could also issue loans and create more promissory notes than what was stored on the premises. Hey presto, the first banks.

In the late seventeenth century the Bank of England – established in 1694 to loan money to the skint government – started writing promissory notes, and the era of currencies issued by central banks had begun. The result was rampant inflation as banks created money out of thin air by printing notes. To end that madness, in 1816 the UK became the first government to tie the supply of money to its actual stores of gold.

This gold standard worked well until the First World War, and then the Great Depression led to massive bank failures that scared the public into hoarding so much gold that the system crumbled. In 1931 the UK officially severed the link to the gold standard. The US followed in 1933.

Divorced from gold, a banknote was now simply a promise from the government that this paper had value. This fiat is enshrined into all modern currency, which is hence known as fiat money. The whole system is built on trust: all users of a currency simply have to trust that the government will guarantee the value of their intrinsically worthless bits of paper.

It works, but understandably gives some people the heebie jeebies. One was a still unidentified person (or persons) going by the pseudonym Satoshi Nakamoto. In 2008 Nakamoto invented Bitcoin, a digital currency that aimed to circumvent the problems of government-issued fiat money. Bitcoin requires no trust and no central banks to own it. Bitcoins are 'mined' by computers performing calculations to verify and record earlier transactions: the bitcoins are a kind of payment for this work. Anyone can mine bitcoins, but the calculations and the software required are not trivial.

Ledgerdemain?

At the heart of Bitcoin is the blockchain, an unfalsifiable ledger of every transaction. It makes sure that no bitcoin can be spent twice, counterfeited or stolen. It is updated constantly and simultaneously, and it is available for inspection by anyone. However, it still relies on trust to some extent, since a bitcoin has no intrinsic value other than what other people are willing to exchange for it.

Bitcoin can be hard to get your head around, but so is fiat money. And it – or something similar – has a good shot at becoming the future. Some central banks are exploring digital currencies as a replacement for existing money because the blockchain provides such iron-clad security.

So if your wallet contains bitcoins, congratulations: you are helping to kill off Tinkerbell. We'll miss her at first, but many people think she has it coming.

FUNNY MONEY

All kinds of strange things have been used as money. The commonest was gold and other precious metals. But currencies did not always have intrinsic value. Feathers, beads and shells have all been used. The most famous example of the arbitrary and slightly surreal nature of money is the rai – massive doughnut-shaped stones that until quite recently served as currency on the island of Yap in Micronesia. Ownership could be transferred in a transaction but the stones were rarely moved. Everybody knew where they were and who owned them, even when they fell out of boats and sank to the bottom of the sea.

WHEN DID WE START BURYING THE DEAD?

THE ONLY CERTAIN THING IN LIFE IS THAT IT WILL ONE DAY END. That dreadful knowledge is perhaps the defining feature of the human condition. As far as we know, we are the only species capable of contemplating the prospect of our inevitable demise.

If it is any comfort, at least you'll probably get a decent send-off. Humans are also the only species to perform the elaborate death rituals that we call funerals. The evidence suggests that we've been performing them for at least 100,000 years. And their origin is a topic of morbid fascination for those who research human evolution.

Funerals clearly fall into the category of 'symbolic activity' alongside art, storytelling, religion and the other cultural trappings of humanity. Ceremonies, deliberate burial and the deposition of grave goods clearly require abstract thoughts about life, death and what it all means. And unlike most other forms of symbolic activity, funerals leave lots of physical evidence behind.

Dead as a dodo

As far as most animals are concerned, a dead body is just an inanimate object. Others clearly have a more complex relationship with death. Elephants appear to be fascinated with the bones of dead elephants, and dolphins have been observed spending long periods of time with corpses.

Chimpanzees are also fascinated with the bodies of other chimps and have been described as displaying behaviours resembling grief, vigilance, respect and mourning. Perhaps, say anthropologists, chimps retain primitive behaviours that were also practised by early protohumans, and which we have elaborated into our formal rituals. We will never know for sure, of course. But the fossil and archaeological record contains tantalising hints of how this kind of behaviour evolved into modern funeral rituals.

The earliest signs are very old indeed. In 1975, on a steep grassy hillside in Ethiopia, palaeontologists discovered 13 partial skeletons of our 3.2 million-year-old ancestor *Australopithecus afarensis* – nine adults, two juveniles and two infants – all within touching distance of one another and apparently deposited around the same time. How they got there is a mystery. There is no evidence of a flash flood or similar catastrophe that could have killed them all at once, and no sign that the bones had been chewed by predators. They are, as discoverer Donald Johanson later wrote, 'just hominids littering a hillside'.

The land of the living

One possible explanation is that the bodies were left there deliberately in an act of 'structured abandonment'. That doesn't mean burial, nor anything with symbolic or spiritual meaning. Even so, it represents a significant cognitive advance over what is seen in chimpanzees, which leave their dead where they fall. Perhaps it was the first stirring of something human; a conceptual division between the living and the dead.

Barring new discoveries, it will be impossible to confirm that australopithecines placed their dead in a special place. But by half a million years ago the evidence is much clearer. Sima de los Huesos – the pit of bones – was discovered in the 1980s at the bottom of a shaft in a cave in the Atapuerca Mountains of Spain. It contained the remains of at least 28 archaic humans, most likely *Homo heidelbergensis*, a probable ancestor of both us and Neanderthals.

How did they get there? They may have accidentally fallen down the shaft, but that seems unlikely from the way the bones fractured and the fact that most of the skeletons are adolescent males or young men. The best explanation is that they were deliberately placed at the top of the shaft after death and then gradually slumped in. If so, this is the earliest evidence of 'funerary caching', or the designation of a specific place for the dead.

A similar discovery was made more recently in South Africa, where 1,500 fossilised human bones and teeth of a previously unknown archaic human, *Homo naledi*, were found clustered in a cave. Frustratingly, we don't know how old they are – anything from 3 million to 100,000 years – and how their species relates to us.

We also have no idea what, if anything, these humans understood about mortality. But we do know that funerary caching became increasingly common. From 500,000 years onwards bodies are often found in places where their presence is hard to account for any other way, tucked into fissures and cracks, in hard-to-reach overhangs or at the back of caves.

From funerary caching it is a short conceptual leap to burial – creating artificial niches and fissures to stash the dead. The earliest evidence we have of this is from two caves in Israel, Skhul and Qafzeh, where the skeletons of 100,000-year-old *Homo sapiens* were found in human-made hollows. These burials may also contain grave goods in the form of animal bones, seashells and ochre. There is also evidence of Neanderthal burials from around the same time.

These burials still do not represent a cultural watershed. Only a handful of such sites are known; compared with the number of people who must have died, they are rare. It also seems likely that cremation was practised, but the evidence is lacking.

It was not until about 14,000 years ago that most dead people were buried in what we would recognise as cemeteries. It is probably no coincidence that around the same time people were settling in one place and inventing agriculture and religion. Our ape past was dead and buried, and symbolic culture was alive and kicking.

ANIMAL RITES

Chimpanzees' understanding of death is impossible to know, but in 2010 primatologists got a rare insight when rangers in Gombe national park in Tanzania discovered the body of a female chimp, Malaika, under a tree. A crowd of chimps had gathered around the body. For the next three and a half hours a succession of chimps approached it; others watched from the trees. Some sniffed or groomed the body. Others shook, dragged and beat it. The alpha male threw it into a stream bed. Many made distress calls.

When the body was removed by rangers, several chimps rushed to where it had lain and intensively touched and sniffed the ground. They stayed for 40 minutes, making a chorus of hooting calls before moving off. The last chimp to visit the spot was Malaika's daughter Mambo.

WHAT WAS THE FIRST COOKED MEAL?

BREAKFAST: FIBROUS AND BITTER LEAVES; FRUIT. Lunch: bark; fruit; raw monkey meat and brains. Dinner: grubs; leaves; fruit.

No, not the latest food fad from Hollywood, but the diet of our closest living relatives, the chimpanzees. It is not exactly appetising or varied. We, on the other hand, have thousands of foodstuffs to choose from, and also an incredibly versatile range of techniques for altering their chemical composition through the application of heat. In other words, cooking.

Cooking is ubiquitous in humans. All cultures, from the Inuit of the frozen Arctic to the hunter-gatherers of sub-Saharan Africa, are sustained by food that has been chemically and physically transformed by heat. It was an incredible invention. Cooking makes food more digestible and kills off the bacteria that cause food poisoning. But where and when it started is hotly debated. You might call it a food fight.

No food without fire

Cooking cannot happen without fire, so the answer might be found by looking for evidence of the control of flames. This is an incendiary topic, as fire is a tricky thing to identify in the archaeological record. The evidence has literally gone up in smoke, and the remains of a deliberately lit fire are hard to distinguish from those of a natural one caused by lightning. This is why archaeologists look for signs of fire in caves.

Traces of ash found in the Wonderwerk cave in South Africa suggest that hominins were controlling fire at least 1 million years ago, the time of our direct ancestor *Homo erectus*. Burnt bone fragments also found at this site suggest that *Homo erectus* was cooking meat. However, the oldest remains of obvious hearths are just 400,000 years old.

The Neanderthals who evolved from *Homo erectus* some 250,000 years ago certainly created fires, as hearths have been found at many Neanderthal sites, some containing burnt bones. We also know from analysing their dental plaque that Neanderthals spiced up their diets with herbs. But we don't know whether they habitually cooked their food.

The earliest firm evidence that our own species was cooking dates back just 20,000 years, when the first pots were made in China. The scorch marks and soot on their outer surfaces point to their use as cooking utensils. But all in all, archaeological evidence doesn't paint a clear picture. We need to look elsewhere.

Around 1.9 million years ago some major changes occurred in hominin biology. Compared with its ancestors, *Homo erectus* had very small teeth, a small body and a much larger brain. According to a controversial hypothesis put forward by primatologist Richard Wrangham, these changes were driven by cooked food. In fact, Wrangham believes that cooking drove our lineage's divergence from more ape-like ancestors and that the bodies of *Homo sapiens* couldn't exist without cooked food.

To understand why, imagine eating the same diet as a chimpanzee. To gain enough calories to fuel your energy-guzzling brain, you would have to devote almost all of your daylight hours to searching for food. Chimps forage more or less continuously; gorillas and orang-utans eat for nine hours a day.

Weak jaw

We would probably have to eat for even longer. Our brains are more than twice as big, and our intestines are far too small to retain low-quality raw food long enough to digest it properly. In fact, our guts are just 60 per cent of the weight expected if we were a great ape of similar stature.

Our small teeth and jaws tell a similar story. They are too small for the task of grinding down large quantities of tough raw food. Compared with earlier hominins such as *Homo habilis*, modern humans, Neanderthals and *Homo erectus* all have small teeth relative to their body size. To Wrangham, these morphological features are adaptations to cooking that arose around 1.9 million years ago.

Cooking certainly changed our ancestors' lives for the better. Heat makes food softer, so less time is needed for chewing. It also releases more calories. Mice fed cooked food get fatter than those fed equivalent raw calories. Heat-treated food is also safer. Scavenged meat has high levels of pathogens. Roasting it on hot coals kills off germs that cause food poisoning. Another benefit of cooking is that it makes otherwise inedible foods, such as tubers, edible. And it frees up time to do more interesting things than just finding food and eating.

Food usually tastes nicer when cooked. We cannot know if our ancestors appreciated the difference, but studies with apes found that they prefer their food cooked, choosing baked potatoes, carrots and sweet potatoes over raw ones most of the time.

Don't eat it all at once

Cooking requires cognitive skills that go beyond controlling fire, such as the ability to resist the temptation to scoff the ingredients, patience, memory and an understanding of the transformation process. Recent experiments with chimps found that they have many of the cognitive and behavioural skills needed for cooking – and therefore it's likely that *Homo erectus* did too.

There are, however, flaws in the cooking hypothesis. Many of the adaptations attributed to cooked food such as large brains could have arisen through an increase in raw meat consumption. The disconnect in time between the biological evidence and the control of fire is another stumbling block.

But whenever cooking was invented, it has evolved into one of the most varied and inventive elements of human culture. We cook thousands of different types of animal, plant, fungus and algae using a dazzling array of techniques. We spend far more hours planning and preparing food than actually eating it, and then sit down to watch programmes about it, hosted by people who have become millionaire household names. We cook, therefore we are.

BROWNED OFF

One of the most important processes in cooking is the Maillard reaction, named after the French chemist who described it in 1912. A reaction between sugars and amino acids, it is what creates the brown compounds that make meat, toast, biscuits and fried foods so delicious. Humans generally prefer food that has undergone the Maillard reaction.

From an evolutionary perspective this is hard to explain. The Maillard reaction makes food – especially meat – less digestible, destroys nutrients and produces carcinogenic chemicals. It may be that the other benefits of cooking food massively outweigh these detriments, and so we have evolved to prefer browned food. But that doesn't explain why it is also preferred by great apes, which can't cook and won't cook.

HOW DID WE TAME THE ANIMALS?

VISIT THE FAMOUS CAVE PAINTINGS AT LASCAUX IN FRANCE AND YOU'LL BE CAST INTO A PREHISTORIC BESTIARY. Of nearly 2,000 images, around half are of animals: horses, stags, bison, felines, bears, birds and rhinos. The people who painted these images 17,300 years ago were clearly obsessed with the animals in their environment: there are no images of landscapes or plants; the rest are human figures or abstract symbols. But to them, the idea of taming or owning a live animal would have been totally alien. The Lascaux animals were wild.

Fast forward to today and our relationship with animals has utterly changed. Over the past 15 millennia we have domesticated hundreds of species of wild animal, from lowly fruit flies to mighty elephants, to serve us in various ways: for food, labour, transport, protection, materials, fertiliser, pest control, herding, companionship, amusement and research.

If you just count livestock, there are about 32 billion domestic animals on Earth. There are maybe a billion dogs and twice as many cats. Beyond that the numbers get hazy, but consider all the rats, mice, guinea pigs, fish, bees, silkworms, escargots, frogs, snakes, medicinal leeches, fruit flies and so on, and you get the picture. Where do they all come from?

For about 95 per cent of our time on Earth, humans lived like the cave painters. Then, somewhere between 30,000 and 15,000 years ago, we started making friends with animals. The first into the fold was an animal that by rights we should have feared and loathed. The result was an altogether tamer animal we still consider our best friend: the domestic dog.

The start of a beautiful friendship

Where, when and how this happened is a matter of debate. The oldest wolf fossils with dog-like traits are from Europe and Siberia more than 30,000 years ago, but DNA evidence points to a more eastern origin about 15,000 years ago.

Wherever and however it happened, there's also the question: Why? One possibility is that it occurred naturally. Wild wolves scavenged from carcasses left behind by hunter-gatherers; some wolves learned to follow the humans and eventually formed a mutually beneficial relationship, helping to track and kill prey and providing protection. Following bands of humans could also have isolated the wolves from their wilder brethren, meaning that they bred only with other human-friendly wolves. The result would have been selection for traits that were beneficial for cohabiting with humans, such as docility.

Another possibility is that humans deliberately adopted wolf pups and bred them – though this seems unlikely given that a wolf eats 5 kilos of meat a day, making it an expensive luxury.

With an oink oink here

Dogs were ideally suited to ice age hunter-gatherer lifestyles, but it wasn't until about 11,500 years ago, when people started to settle in villages, that domestication of animals began in earnest.

Pigs came first, with sheep, goats and cattle following soon after. These animals were almost certainly domesticated for their meat, milk, wool, horns and hides, though full domestication was probably preceded by thousands of years of increasingly hands-on management of their wild relatives. There are also suggestions that cattle were domesticated by shamans for ceremonial purposes and only later repurposed as walking larders.

Domestic cats also seem to have their origins around this time, though pinning down the details is like herding cats. Their wild ancestor is the Near Eastern wildcat *Felis silvestris lybica*, which was probably attracted to human settlements by the rodents that massed around grain stores. People would have noticed their utility as pest controllers and also their adorable cuteness. Thus began the slightly stand-offish relationship that persists today. Cats are still not considered fully domesticated – they retain much of their wild nature and, unlike almost all other domesticated animals, do as they please.

The Fertile Crescent was the trailblazer, but there were at least five other places where animals were domesticated in prehistoric times. Each made a unique contribution to our growing menagerie: horses, ducks and silkworms from China; water buffalo from India; donkeys and dromedaries from Africa; turkeys from Mesoamerica and guinea pigs and llamas from South America.

How did the chicken cross the world?

Perhaps the most significant domestication outside the Fertile Crescent was the red jungle fowl, which joined the fold across east and south Asia around 7,000 years ago. The resulting species is the common-or-garden chicken, now the most numerous livestock animal in the world. Poultry farming produces about 40 billion chickens a year.

This is all the more remarkable given that jungle fowl are home birds that do not migrate, are poor flyers and occupy a small home range. Their distribution on every continent except Antarctica can be entirely attributed to humans. In fact chickens are so intimately connected to us that their DNA has been used as a proxy to reconstruct how humans colonised the expanses of the Pacific Ocean.

DNA suggests that the chicken run began at least 3,000 years ago, both westwards towards central Asia and eastwards into Polynesia and beyond. Chickens entered Africa via Egypt around 1200 BC; centuries later the Romans spread them around their European empire. This global pincer movement eventually converged in the Americas, where chickens were introduced first by Polynesians and later by Europeans and Africans.

Humans have continued to subjugate animals for various purposes. Some of the more recent additions are ferrets, derived from the European polecat about 1500 BC; goldfish from Prussian carp in China around AD 300; and the common fruit fly *Drosophila melanogaster*, which was adopted as a model organism for genetic research in about 1910.

FLOPPY EARS AND CUTE FACES

Charles Darwin saw selective breeding of domestic animals as an analogy for natural selection in the wild. He was also the first to notice something odd: all domesticated mammals possess a similar suite of biological differences from their wild relatives, known as the domestication syndrome. They include docility, floppy ears, a curly tail, reduced size, a juvenile-looking face and altered coat colour. Many of these traits are obviously useful (or at least desirable) to humans and were probably selected during domestication and further selective breeding: others seem to have come along for the ride, possibly because they are controlled by the same genes.

WHEN DID WE START WORSHIPPING GODS?

THE HADZA PEOPLE OF TANZANIA DO NOT CONCERN THEMSELVES MUCH WITH GODS. They have origin myths and supernatural stories, but their belief systems are informal and their deities distant, impersonal and unconcerned with everyday morality.

The Hadza and other hunter-gatherer groups are often held up as models of how our distant ancestors lived. If so, they lacked something that has been a central part of human life for most of recorded history: organised religion.

Even in these increasingly secular times most people identify with one or other of the major world religions: Christianity, Islam, Hinduism, Buddhism, and the rest. Unlike the Hadza's informal folk religion, these are characterised by doctrine, prescribed rituals and a hierarchical power structure. They are among the most powerful driving forces of human history, for both good and ill. Where did they come from?

Born believers

To begin to answer that question, we first need to ask a different question: Why are people religious at all? For many the answer is obvious: Because god exists. Whether or not that is true, it tells us something interesting about the nature of religious belief. To most people, belief in god is effortless, like being able to breathe. In recent years scientists have produced an explanation of why this is so. Called cognitive by-product theory, it holds that humans are 'born believers'. Our brains are naturally inclined to find religious explanations appealing and plausible.

For example, evolution has endowed us with a default assumption that everything in our environment is caused by a sentient being. That makes evolutionary sense: our early ancestors were regularly attacked by predators, and when any rustle in the bushes might be a threat, it was better to err on the side of caution. But it also primes us to see agency where there is none; to assume that the world we see around us was created by somebody or something, that everything has a purpose, and that effects are always preceded by causes. This, of course, forms the core tenet of most religions: that an unseen agent is responsible for doing and creating things in the world.

This and other inbuilt cognitive tendencies make humans gravitate towards supernatural beliefs like those of the Hadza. But they don't fully explain the origins of big, organised religions. The answer to that may lie on a hilltop in Turkey amid the ruins of what is widely seen as the world's oldest temple. Discovered in the 1990s, Göbekli Tepe is a labyrinth of circular stone enclosures measuring up to 30 metres across. In the centre are pairs of 6-metre pillars encircled by smaller T-shaped statues. Some are engraved with belts and robes, while others bear grotesque carvings of snakes, scorpions and hyenas.

It is easy to see why the archaeologists who excavated Göbekli Tepe interpreted it as a religious complex. But to make that idea stick, they had to challenge a strict orthodoxy about the origin of organised religion: that it was one of the products of the Neolithic revolution.

Village people

In a nutshell, this holds that around 10,000 years ago humans began to abandon the nomadic lifestyle and settle in permanent agrarian communities. By about 8,300 years ago, people in the Levant had the full package of Neolithic technologies: agriculture, domesticated animals, pottery and settled villages. They are also assumed to have had organised religion. Indeed, the invention of such religions is seen as being crucial to their success. Before the transition, humans lived in small and intimate family groups. Afterwards, they largely lived among unrelated strangers, requiring unprecedented levels of trust and cooperation.

In evolutionary biology, trust and cooperation are usually explained in one of two ways: kin helping one another, and reciprocal altruism, or 'you scratch my back and I'll scratch yours'. But neither of these easily explains cooperation among large groups of unrelated humans. With ever-greater chances of encountering strangers, opportunities for cooperation among kin decline. Reciprocal altruism also stops paying off.

This is where religion comes in. Many of today's religions encourage cooperation and altruism, and some early variants presumably did the same.

Religions with these features would have helped to bond unrelated people into shared communities, acting as the social glue that held the fragile new societies together. Such groups would have grown larger than neighbouring ones, and outcompeted them for resources. As these groups grew, they took their religions with them. Most of us now are at least vaguely connected to one of these highly successful religions.

The problem with Göbekli Tepe is that it is too old to fit into this story. The oldest buildings date back 11,500 years, to a time when people were still hunting and gathering. No evidence of agriculture has been found at the site and there is also no sign of any kind of permanent settlement. Yet it is clear that Göbekli Tepe was the creation of a sophisticated society, capable of marshalling the labour of large numbers of people. The evidence also strongly suggests that this society had a common system of beliefs and rituals, which they gathered at Göbekli Tepe to enact and share. In other words, the trappings of an organised religion.

The discovery of Göbekli Tepe – and other nearby complexes with similar monuments, some even older – has therefore turned the orthodoxy on its head. Rather than agriculture creating fertile conditions for organised religion to take root, it appears to be the other way round. Ritual gatherings, not farming, initially pulled people together into a larger society. And the need to feed people at these gatherings may have been the impetus that got agriculture going. Tellingly, recent genetic work pinpoints the origin of domestic wheat to a spot very close to Göbekli Tepe.

DOORWAY TO THE AFTERLIFE

We will never know what the people who built and worshipped at Göbekli Tepe believed, but archaeologists have a few ideas. One of the site's most prominent features are door-like 'portal stones', often decorated with images of predators and prey. The holes are big enough to crawl through, suggesting that visitors may have done so to symbolise birth or death. What is more, amidst the debris of the site archaeologists have found many bones, including human remains and a surprising abundance of rooks and crows – birds that are known to be drawn to corpses. This is yet another reason to believe that one of the building's functions was for a death ritual – which, of course, is a universal feature of modern religions.

WHEN DID WE START GETTING DRUNK?

What's your poison? If you're a typical human being, there's only one answer to this question: ethanol. This toxic, intoxicating liquid has – for better and worse – been part of human culture for millennia. Ancient cultures the world over stumbled on the recipe for alcohol, and then stumbled about in a drunken stupor. Only the indigenous people of the Arctic, Tierra del Fuego and Australia failed to discover fermentation. Pretty much anything that can be fermented to make ethanol, we have fermented: grapes, grains, apples, pears, honey, rice, milk, tree sap, nettles, potato peelings. All we need to have a good time is some sugar, some yeast and a bit of patience.

Our fondness for ethanol goes back a long way. The oldest known alcoholic beverage is about as old as farming. But our association with the stuff goes back much, much further than that.

Humans are not the only species that likes a drink. Fruit flies eat fermented fruit, apparently without it affecting their behaviour. Other animals get noticeably drunk: waxwings have been spotted flopping around in trees or crashing into buildings after eating fermented berries. Elephants are notorious for their drunken rampages. In one tragic episode in Assam, India, a herd of elephants stumbled across casks of beer, chugged the lot then trampled at least six people to death.

Among our primate cousins, vervet monkeys on the Caribbean island of St Kitts are notorious for stealing cocktails, while lorises in the Malaysian jungle are habitual drinkers, taking daily nips of the fermented nectar of the bertam palm. Wild chimps in Guinea have been observed getting plastered on palm wine.

This kind of behaviour can be traced back to the evolution of fruiting plants around 130 million years ago, when our ancestors were little more than tree shrews in the shadows of the dinosaurs. Early mammals took advantage of this new source of food, and so did microorganisms. A genus of fruit-loving yeasts called *Saccharomyces* (Greek for sugar fungus) arose and rapidly evolved a sneaky adaptation. Instead of breaking down the fruit sugar completely, they evolved the ability to partially break it down, producing ethanol. This was less efficient at liberating energy but had the advantage of poisoning other microorganisms that also liked to feed on fruit.

Fruity aroma

Saccharomyces would have fed on ripe fruits, so the smell of ethanol was a sign that fruits were ready to eat. Natural selection then favoured fruit-eating mammals that could use the smell of ethanol to locate nutritious food. Those that liked the taste of fermenting fruit would have outcompeted those that didn't, and so a liking for the taste and psychoactive effects of alcohol became part of our primate ancestors' biological make-up. It was love at first swig.

Fast forward a few million years and a liking for alcohol may also have offered benefits once our ancestors began to farm. At the dawn of agriculture, around 10,000 years ago, people in small settlements began to ferment foods and drinks. This would have allowed them to preserve surplus grain, in essence by favouring yeasts in place of food-spoiling bacteria. It would even have made grain more nourishing because fermentation produces nutrients, including B vitamins.

Fermentation also offered a way to sterilise liquids. In the unsanitary conditions faced by early settled communities, fermented drinks were safer than water. Alcohol consumption might also have helped smooth social interactions, which would have become more complex as communities grew.

As for how we learned to make alcohol, perhaps the first farmers stumbled across the recipe by accident, when stored wheat and barley became contaminated with *Saccharomyces* yeasts. Some anthropologists suspect that we first cultivated cereals for fermentation rather than for food. Beer, it seems, came before bread.

Grapes were domesticated later and, thanks to the fungus that grows on their skin, would have naturally fermented into wine when left in the sun. The oldest remains of actual alcoholic drinks confirm their antiquity. They are found as residues on the insides of 9,000-year-old pots from a site called Jiahu next to the Yellow River in China. Chemical tests show that the residues contain a fermented mixture of rice, hawthorn fruit, grapes and honey.

Old wine in old bottles

Pottery fragments have also yielded residues of the oldest wine, containing a mixture of fermented grapes and resin from the 7,000-year-old archaeological site of Hajji Firuz Tepe in Iran. The earliest chemical evidence of beer comes from the same region about 1,500 years later. Cocktails were also on the menu: 3,000 years ago Bronze Age Greeks were drinking a mixture of wine, beer and mead. Many of these ancient drinks have since been recreated by biomolecular archaeologists. Some are said to be quite tasty.

However fermentation was discovered, people quickly realised that it was a gift that kept on giving. You could do it over and over again simply by taking a sample from one live batch and using it to inoculate a new one.

Brewer's yeast changed many times as agriculture spread and different human cultures emerged. New forms emerged in association with brewing and wine-making in different regions. Some were repurposed as bread yeasts.

Fermentation can only take you so far, however. Eventually the yeast is poisoned by its own waste. The strongest fermented drinks are only about 15 per cent ethanol by volume. However, the invention of distillation in China about 1,000 years ago solved that problem, allowing fermented drinks to be converted into hard liquor. Cheers!

CHOCAHOLICS ANONYMOUS

Even teetotallers have a reason to toast the invention of booze: chocolate. Mesoamericans – who flourished in Central America before the Spanish arrived – developed it as a by-product of making alcohol.

Chocolate's flavour develops only when the watery pulp of raw cacao fruit and seeds are fermented together. It is not obvious how the Mesoamericans discovered this process unless they were fermenting the fruit for another reason – which they were, to brew a beverage called chicha. The key step probably came when the chicha brewers ground up cacao seeds left after fermentation, added them to thicken the beer, and found that they imparted a pleasing chocolatey taste.

WHY DO WE NEED SO MUCH STUFF?

IN ONE OF HIS MORE IDEALISTIC MOMENTS, JOHN LENNON ASKED US TO IMAGINE LIFE WITHOUT POSSESSIONS. Give it a go: it's not easy, is it? In fact, it is almost unimaginable. Without clothes, a roof over our heads, some means of cooking and a supply of clean water, we could barely survive. Try to imagine life without a bed, a bath, towels, light bulbs and soap – let alone indulgences and luxuries, and all those objects with sentimental value. Possessions define us as a species; a life without them would be barely recognisable as human.

Our closest living relatives make do with none of this. Chimps employ crude tools and build sleeping nests, but abandon them after one use. Most other animals get by without possessions at all.

How did we evolve from unencumbered ape to hoarding human? Answering this question is not easy. For one thing, drawing a line between 'possessions' and 'non-possessions' is not straightforward: do you own the soil in your garden, for example, or the water in your taps? And when you discard something, when does it cease to be yours? What's more, many objects our ancestors may have owned – animal pelts or wooden tools – are unlikely to have survived in the archaeological record.

Nonetheless, there are clues about humanity's first possessions. The earliest stone tools, made some 3.3 million years ago, are an obvious place to start. They were designed to do a job, and must have been held by an individual for a time. Yet they were extremely simple and expendable, like chimpanzee tools.

That's mine!

But as tools became more sophisticated and harder to make, a sense of ownership must have started to evolve. Tools became the first real possessions – items that were valued by their owner, in their possession for a length of time and recognised by (possibly envious) others as belonging to somebody else. The concept of ownership really took off with the advent of spear- and arrowheads, which first appeared in Africa at least 300,000 years ago. Hunters would have retrieved them from kills, and used them again and again.

Another key early possession was probably fire. Some contemporary hunter-gatherer groups carry embers around with them, and so can be thought of as 'possessing' fire. Our ancestors may have done the same. The earliest convincing evidence of controlled use of fire dates to around 800,000 years ago.

Clothing, too, made an early entrance. Genetic evidence from body lice suggests we started wrapping ourselves up about 70,000 years ago.

Once we possessed fire, clothing and sophisticated tools, we presumably came to depend on them for survival – especially after colonising colder climates. Our belongings started to become part of our 'extended phenotype', as crucial to survival as a dam to a beaver.

Conspicuous consumption

With time, there was another leap forward. Objects became valued not only for their utility but also as prestige goods to advertise the skill or social status of their owner. Eventually, certain objects became valued for these reasons alone – jewellery, for example. The earliest evidence of this is a small number of 100,000-year-old shell beads found in Israel and Algeria.

It is clear that tens of thousands of years ago the relationship between people and objects had already evolved beyond utility and survival value.

But the amount of stuff that people could accumulate was constrained by their nomadic lifestyle, leading some archaeologists to speculate that bags or papooses might have been among our earliest and most transformative possessions. Bags allow people to accumulate more things than they have hands, and to carry them around. Unfortunately bags are made of biodegradable material, so we have no idea when they were invented. The oldest we know of are about 4,000 years old.

This changed with the switch to a settled lifestyle. Once people chose to live in one place, their possessions began to accumulate. This lifestyle also heralded a new form of society and economy. Groups became larger and hierarchies developed, with the status of important individuals bolstered by prestige possessions such as fine clothes and jewellery. In fact, some archaeologists argue that societies could not have become complex and hierarchical without an associated 'material culture'.

Grease my palm

This switch to sedentarism drove materialism in another way. When people settled down, they became more susceptible to environmental disaster, which fuelled an urge to accumulate possessions as an insurance policy. Another insurance policy was to develop relationships with neighbouring groups. The exchange of non-necessary goods could grease those relationships. Eventually, when societies became even larger and more complex, material goods became a store of wealth. Trade in such goods eventually led to the development of money.

There are a number of groups in the world today who don't live in large, complex societies, and who have very few possessions. The Hadza people of Tanzania, for example, have few material goods and a culture of enforced sharing. But the vast majority of people don't live like this and, as a consequence, are surrounded by stuff and are in possession of a seemingly insatiable desire for more.

So what are the chances of breaking the human habit of owning too much? When you consider our reliance on things to survive and signal social status, it doesn't seem likely. Dream on.

DOG OWNERS

In 1776 philosopher Adam Smith noted a curious fact about animals: they don't appear to own things. 'Nobody ever saw a dog make a fair and deliberate exchange of one bone for another with another dog.' In many respects he was right. Only humans have a complex system of property. But some animals arguably do have possessions: birds have nests and beavers have dams. Squirrels and scrub jays cache food. Bowerbirds collect shiny and colourful objects to attract mates. And many animals defend a territory.

But none of these behaviours comes close to the sophistication of human ownership. The reason is simple: language. Without words, mutually understood rules and institutions to enforce them cannot exist. Possession is nine-tenths of the law – and the law is made of words.

WHEN DID WE SWAP OUR FUR FOR CLOTHES?

HUMANS DIFFER FROM THE OTHER GREAT APES IN MANY RESPECTS, BUT ONE OF THE STARKEST IS WHAT WE LOOK LIKE NAKED. Chimps, bonobos, gorillas and orangs are almost entirely hairy. We are almost entirely not.

Of course, we rarely see one another naked, and so tend not to notice how hairless we are. That is because we habitually wear clothes, as if losing the hair that covered our early ancestors was a big mistake.

In some respects it was. Humans evolved in Africa where keeping cool was a bigger challenge than keeping warm. Hairlessness would have been an advantage in this environment, especially given that our cooling system is sweating, which doesn't work well when you are covered in fur. But once we became hairless, our horizons narrowed. Head too far north or south and the environment would have been too harsh to survive.

That is clearly no longer a problem. Modern humans are distributed right across the globe. One of the technologies that enabled this conquest was clothing (fire and shelter helped a lot too, but are not so easy to carry around as a jumper). Clothing also serves purposes beyond keeping warm and dry. It is an important social signal, transmitting messages about who we are.

Wrap up warm

The origin of clothing, then, was an important event in human prehistory. Unfortunately it is also shrouded in mystery. Clothing is made of biodegradable material – wool, fur, leather, plant fibres – and does not readily survive the ravages of time. The oldest known items of clothing are just a few thousand years old, but we know that humans must have been wearing them much earlier: the Siberians who walked into Alaska across the Beringia land bridge 15,000 years ago – the peak of the last ice age – must have been clothed. Going further back, it is hard to imagine people colonising Europe 40,000 years ago without something to wrap up warm in.

There is proxy archaeological evidence which tells us that clothing is probably an ancient invention. Cave paintings in France that show people wearing clothes have been dated to 15,000 years ago, but their authenticity is in doubt. The oldest sewing needles are about 40,000 years old, and scrapers for preparing animal hides date back half a million years, but both may have been used for purposes other than making clothes. As for when our ancestors became hairless, there is even less solid evidence.

Fortunately, we don't need archaeology. Both getting naked and then getting dressed have been dated from a rather unlikely source: lice. For most species of mammal, these are an irritating fact of life. Most primates are infested with a single, specialist species, though orang-utans and gibbons somehow escaped. But we humans, with our patchy hair and our clothes, are blessed with three kinds: head lice, pubic lice and clothes lice, all of which eat blood. We are truly the lousiest of the apes. On the upside, they have stories to tell about our past.

The first concerns hairlessness. We, too, presumably once had a single species of louse occupying the hair that stretched from head to toe in one continuous habitat. This turns out to be the ancestor of the modern head louse, *Pediculus humanus*. As we lost most of our body hair, its habitat shrank. But a new one opened up: pubic hair. This is coarser than head hair and too thick for the feeble and delicate head louse to cling to. The species that took advantage is *Pthirus pubis,* the pubic louse. They are much bigger and more robust than head lice (hence their nick-name, crabs). They also like to live in facial hair, eyebrows, armpit and chest hair and, occasionally, head hair.

You might assume that pubic lice evolved from head lice, but they did not: their closest living relative is the gorilla louse, *Pthirus gorillae.* At some point in the past, this species jumped from gorillas to humans. Exactly how is a delicate topic: we don't need to go there. But genetics tells us that the two species diverged about 3.3 million years ago, suggesting that our ancestors had separate head and pubic hair by then. This is staggeringly early, long before *Homo sapiens* evolved. We have always been hairless.

Threads of evidence

What about clothes? When prehistoric humans began covering their nakedness, they created yet another new niche for lice. This time the head louse got in first. Clothing lice resemble bigger, tougher head lice. There is no doubt the two are closely related.

But they are different enough to use genetics to put a date on their divergence. A recent analysis concluded that a common ancestor of the two lived at least 83,000 but possibly as much as 170,000 years ago. It seems our ancestors began wearing clothes before they started migrating out of Africa to occupy the world. If so, it is tempting to speculate that clothing was one of the technological breakthroughs that enabled them to do so.

Exactly what the first clothes looked like or were made of is anyone's guess. But their invention coincided with the first stirrings of symbolic culture, so it is reasonable to assume that functionality was soon joined by flamboyance and fashion. Nice furs – shame about the lice.

BOOTED AND SUITED

The oldest intact shoe dates from about 5,500 years ago. Found in a cave in Armenia, it is made from a single piece of cowhide laced with a leather cord. Shoes must have been invented earlier, however, and indirect evidence backs this up. Forty-thousand-year-old toe bones from a cave in China show signs that their owner was a habitual shoe-wearer.

Archaeologists measured the shape and density of the bones and compared them with those from twentieth-century Americans, late-prehistoric Inuits and other late-prehistoric Native Americans. Shoes alter the way we walk. The toes curl far less and so the bones experience less force, leading to telltale anatomical differences. Modern shoe-wearers have wimpy little toes, whereas barefoot native Americans had strong, large ones. Shoe-wearing Inuits lie somewhere in between. The Chinese bones most resembled the Inuits', indicating that their owner regularly wore shoes.

WHAT DID THE FIRST MUSIC SOUND LIKE?

Deep in the Congo rainforest live some of the most musical people on earth. The Mbenzélé are hunter-gatherers who rarely leave the forest and don't have radios or TVs. Their contact with outsiders is minimal, and yet their musical compositions – for voice, handclaps and drums – have been likened to the most sophisticated symphonic music, with extraordinary harmonies and polyrhythms.

The Mbenzélé are among the best evidence we have that music is part of human nature. Like language and religion, all cultures have it. It is one of the things that makes us human, and also one of the hardest to explain.

Charles Darwin, who had something perceptive to say about the origin of most human traits, described music as among our 'most mysterious' faculties. He suggested it began as a 'musical protolanguage' – a kind of vocal mating display akin to birdsong – that eventually diverged into two separate traits: music and speech.

Others have suggested that music is a kind of brain training that sharpens various mental skills such as memory and emotion.

It has also been described as 'auditory cheesecake', a pleasurable experience that just happens to appeal to us because of other mental traits such as pattern recognition. Unfortunately, there is little evidence for any of these ideas.

Finding our voices

One reason is that the origins of music are lost in time. The oldest musical instruments are the remains of bone flutes found in European caves also containing paintings. They date from between 42,000 and 15,000 years ago, a time when creativity was flowering.

One of the oldest, the Divje Babe flute, was found in what is now Slovenia among the remains of Neanderthals. Some believe it was made and used by them, which would be interesting – if Neanderthals were musical, that suggests our common ancestor was too, which pushes the origins of music back to at least half a million years ago. But we have no physical evidence, which is no surprise because the earliest music was presumably sung.

One promising avenue for exploring the deeper origins of music is animals. On the whole there is not much to explore: most animals are indifferent. Given a choice between music and silence, monkeys choose silence every time. They also do not seem to discriminate between tunes and discord. In this respect they resemble people with amusia, a rare neurological deficit that robs them of what is normally an instinctive appreciation of music. To these unfortunate people, music is tedious noise and one piece sounds very much like another; a few cannot even distinguish between music and somebody hitting a drainpipe with a spanner. That may be because they lack a brain system that makes us receptive to music, which hints that music is uniquely human.

Bird songs

The animal kingdom is not completely tone deaf, however. Many birds sing complex songs and can learn new ones. Java sparrows can distinguish between different styles of music: they prefer the melodies of Bach to the atonality of Schoenberg and choose to listen to Bach rather than silence.

But that does not make them fully musical. Birdsong serves a strict function. It is largely confined to males, and its purpose is to woo a mate and protect territory. In contrast, we make and listen to music for all sorts of reasons – sex, bonding and territory are among them, but we also do it to enhance or change our moods, to worship gods, to motivate ourselves, to help us concentrate, or just for pure pleasure.

Humans also deliberately compose new music, invent new styles and seek out new musical experiences. Even the most flexible bird singers stick to a core repertoire. The prevailing view is that music is unique to our lineage – which doesn't tell us much about its origins.

But there is one feature of music that seems to go deeper into the evolutionary past: rhythm. Unlike pitch and harmony, it shows commonality across all musical cultures. Human babies respond to rhythm long before they develop an appreciation of harmony. And many non-human animals are moved by it.

Our closest living relatives are rhythmic. Wild chimps drum on resonant objects such as the buttresses of trees. And when researchers gave bonobos at Jacksonville Zoo a drum kit, they quickly learned to play it.

Many of these animals are social species, which some see as a clue. There may be a link between a sense of rhythm and social behaviour – especially the need to coordinate actions. Rhythm acts as a kind of social glue. Bonobos and chimps both live in large groups that require individuals to evaluate and respond to the actions of others. Perhaps predicting the timing of group movements, and synchronising actions with those of others, strengthens the neural circuits involved in rhythm.

Eat to the beat

Our ancestors would also have needed social coordination for activities such as tool-making, hunting and preparing food. Over time, such behaviours may have become overtly rhythmic, since working to a beat helps people to coordinate their actions – just look at work songs and sea shanties.

Repetitive, rhythmic movements may also have developed to help groups bond, deepening the social ties that are crucial for human survival.

If this model is correct, then our musicality started out as a sense of rhythm. Over the millennia it has expanded into something far more sophisticated than that of any other animal. We may never understand how or why, but that shouldn't stop us from enjoying its unique pleasures. Brain scans show that music activates the same parts of the brain that respond to food, sex and drugs. Rock and roll!

BLOCK-ROCKIN' BEASTS

Many animals can move in time to a beat. Perhaps the most famous is Snowball, a sulphur-crested cockatoo who became famous in 2009 when a YouTube video went viral. It shows him dancing exuberantly – and in perfect time – to a song by the Backstreet Boys.

When researchers searched the internet for more, they found 14 other types of animal with a developed sense of rhythm, including macaws, parakeets, Asian elephants and a young sea lion called Ronan. A further 500 videos showed animals such as dogs, ducks and owls moving to music, though failing to keep time. Many humans also possess this trait, which is known as 'dad dancing'.

WHO INVENTED TOILET PAPER?

THE 1850S WAS A GOLDEN DECADE FOR HOUSEHOLD CLEANLINESS. It witnessed the birth of both the dishwasher and the washing machine. But neither invention was quite as revolutionary as that of Joseph C. Gayetty of New York City. In an advert in *Scientific American*, he declared it to be a 'grand and unapproachable discovery' and 'the greatest blessing of our age'. The small print revealed what it was: Gayetty's Medicated Paper, America's first commercial toilet paper.

Gayetty's announcement proved to be surprisingly provocative. Loo roll may now be considered an essential home comfort, but in the 1850s the idea of paying good money for mere 'bum fodder' was greeted by a chorus of mocking laughter. What was wrong with the corn husks and pages torn from newspapers, magazines and catalogues that had served so well and cost so little? Some catalogue publishers in the US had even started piercing a hole in the corner, as if tacitly accepting that their pages were destined to be hung in a latrine and used as toilet paper.

Bummed out

Gayetty's paper hit an especially bum note among medics. According to Richard Smyth, the author of *Bum Fodder: An absorbing history of toilet paper,* they were particularly concerned by the assertion that the new paper could cure piles, and soon took to the pages of leading medical journals to complain.

Despite his grandiose claims, Gayetty was not the first to invent toilet paper. The Chinese had got there hundreds of years earlier. Paper had been circulating in China since the second century AD and it didn't take long for people to stop reading and start wiping. Even the Emperor Hongwu, a brutal despot ruling in the fourteenth century, showed his sensitive side by ordering 15,000 sheets of extra-soft, perfumed toilet paper for his imperial household.

The Chinese also seem to have been first with another essential tool of personal hygiene, the toothbrush. Many ancient cultures used chew sticks to keep their teeth clean – in fact, all civilised people seem to have used some sort of instrument for dental hygiene – but it wasn't until the fifteenth century, during the Ming Dynasty, that actual brushes appear. They were made from coarse pig bristles attached to a wooden or bone handle. European travellers to China brought toothbrushes home with them, and the technology spread to the west.

Toothpaste was an even earlier invention. The ancient Egyptians, Romans and Greeks used various substances to keep their teeth clean, though the ingredients were rather basic and abrasive: ashes, eggshells, pumice, powdered charcoal, tree bark, salt, crushed bones and oyster shells have all been found. Soap – invented in

Babylon around 2800 BC – was another common ingredient. The Romans added flavouring to help with bad breath. The Chinese appear to have invented the first minty toothpaste long before the toothbrush came about.

A fistful of leaves

The Chinese preference for toilet paper, however, did not travel well. The people of Britain were content with fistfuls of wool or leaves. Aristocrats would deploy scraps of linen. Or rather, they'd have someone deploy them on their behalf: a servant's manual from the fourteenth century advises the 'groom of the stool' to be ready with an 'arse-wipe' at the critical moment.

With the advance of the printing press, people soon turned to the disused pages of pamphlets and books. As the seventeenth-century author Thomas Browne wrote: 'He that writes abundance of books, and gets abundance of children, may in some sense be said to be a benefactor to the public, because he furnishes it with bumfodder and soldiers.'

Gayetty was not alone in his attempt to commercialise toilet paper. But it was his product that caused the biggest storm. The sheets, Gayetty declared, were 'delicate as a bank-note and as stout as foolscap'. But what really riled the medical establishment was his claim that printer's ink was poisonous and caused haemorrhoids, and that his paper could 'cure and prevent piles'. There's no truth in the claim, but that did not stop many companies from pushing loo roll as a remedy until the 1930s.

Medical journals soon went on the attack. The *New Orleans*

Medical News and Hospital Gazette declared: 'Mr Gayetty of New York City has found that the public mind is prepared for anything whatever in the shape of humbuggery.' The *Medical and Surgical Reporter* also accused Gayetty of taking advantage of the public, drolly saying that he was attempting to 'catch them with their breeches down.' *The Lancet* was less worried about the general public than the fate of the surgeons who made a good living curing piles. 'Their occupation is now gone to the wall. All that is required is a simple piece of paper with the name "Gayetty" stamped on it.'

Where there's muck there's brass

But even if it didn't cure piles, the public appreciated the comfort of toilet tissue, and it soon spawned a host of me-too products. However, consumer expectation does not seem to have been high. In the 1930s Northern Tissue was able to make a selling point of the fact that its paper was 'splinter free!' Today, the toilet paper industry is worth $3.5 billion annually in the US alone, with the average individual working their way through more than 20,000 sheets a year. Add that to the $3 billion spent on toothpaste and mouthwash, and it is clear that personal hygiene is big business at both ends of the alimentary canal.

SMELLS LIKE TEEN SPIRIT

The human armpit, known to anatomists as the axilla, seems almost specifically designed to be malodorous. Unlike other regions of skin, which are classified as dry, moist or sebaceous (greasy), the armpit is both moist and sebaceous at the same time due to a high density of sweat and sebaceous glands. It also has lots of apocrine, or scent, glands, which pump out a mixture of proteins and lipids. This moist, rich environment is perfect for bacteria, which colonise the skin, dine out on sebum and other nutrients and excrete foul-smelling waste otherwise known as BO.

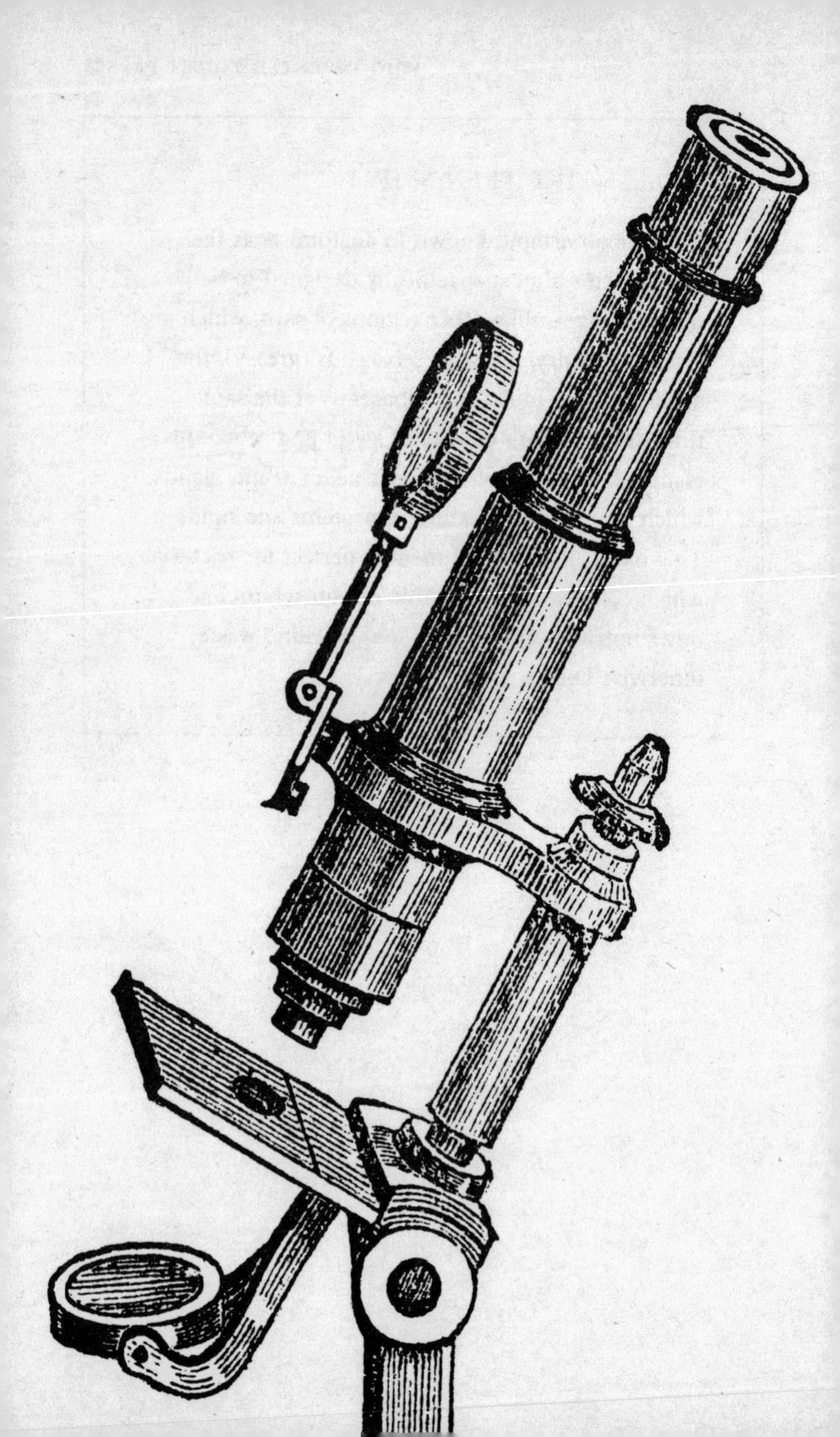

PART FIVE

KNOWLEDGE

WHEN DID WE START WRITING?

IF YOU WANT TO RECORD EVERY THOUGHT THAT HUMAN BEINGS HAVE EVER HAD, HERE'S WHAT YOU NEED: 26 ABSTRACT SHAPES, SOME DOTS, SQUIGGLES AND DASHES AND SOME SPACES. That's all.

Writing is one of the greatest inventions in human history – perhaps the greatest, since it made history possible. Before writing, all ideas were transient, passed on orally or doomed to die. Writing allowed them to be encoded independently of mortal human brains, and built on. As linguist Peter Daniels wrote in the 1996 book *The World's Writing Systems*: 'Humankind is defined by language; but civilisation is defined by writing.'

On the scale of human existence, writing is a recent invention. Humans have spoken languages for at least 100,000 years, but it was only around 35,000 years ago that the rudiments of writing emerged.

These early signs are found in the famous painted caves of Palaeolithic Europe. Among the animals are 26 abstract symbols including geometric shapes, zig-zags, arrows and clusters of dots, drawn in a consistent style. They recur in 146 caves in France dating up to 10,000 years ago, and some are found at other sites across the world.

The writing on the wall

What they meant, if anything, is lost, but to the researchers who study them they contain tantalising hints of writing. Certain signs frequently appear in pairs – a feature typical of early writing systems, where the combined symbols represent a new concept. Others perhaps represented part of a larger figure. W shapes found in Chauvet cave appear to be the tusks of a mammoth, without the body. This feature, known as synecdoche, is common in primitive writing systems, which use pictures to represent objects and ideas.

These cave symbols aside, the earliest known pictographic scripts date to the Neolithic, a time of explosive cultural innovation. One is a collection of 16 symbols engraved on tortoise shells found in graves at Jiahu, China, dating back about 8,500 years. There is also the Vinča script – dozens of recurring symbols inscribed on hundreds of artefacts found across south-eastern Europe. Both scripts look like writing, but they are undeciphered so no one knows for sure.

The earliest true writing capable of recording the full complexity of spoken language emerged in the cities of Sumeria 5,300 years ago. Inscribed on to clay tablets using a blunt reed, cuneiform started as an accounting system to document things like the beer rations paid to workers – a necessary invention for a society that was growing ever more complex. Originally consisting of pictograms – a jar to represent a beer, say – by around 4,600 years ago the signs had evolved to represent syllables, and so could be used to write down the language. The word for arrow was *ti*, for example, so the pictogram of an arrow came to represent the syllable *ti* in words like *til*, which meant 'life'. This type of script is called a syllabary; some modern languages including Japanese are written this way.

Know your ABCs

Two other writing systems emerged around the same time: Egyptian hieroglyphs and the Indus script in what is now India and Pakistan. Hieroglyphs are largely logographic, meaning that each symbol represents a complete word. But they also contain the earliest elements of an alphabet, where symbols represent individual sounds.

These are just a handful of the hundreds of different writing systems that have come and gone as civilisations have risen and fallen. The variety is staggering. Besides syllabaries, logographs and alphabets there are also abjads, which, like Arabic, have no vowels.

Some scripts are read from left to right, others from right to left, and some both: a few scripts turn back on themselves at the end of each line and continue in the opposite direction. They are known by the amazing word boustrophedon, which means 'in the manner of an ox turning'.

Our Latin alphabet originated some 4,000 years ago in a script used to record a language in ancient Egypt. It evolved from a pictographic system; the letter A, for example, is derived from an inverted drawing of a bull's head, called an alf.

The script was adopted and adapted by the city-states on the shores of the Mediterranean, and was spread around the region by Phoenician traders. The Phoenician script has 22 letters and was probably the first purely alphabetical system.

Around 3,000 years ago, it was borrowed and adapted by the Greeks. Later, the Romans borrowed it back, dropping some letters and adding new ones. Today the Roman alphabet is used by some 5 billion people, and is the most common of the 35 or so scripts in widespread use around the world. Thanks to it and others, we can access the thoughts and experiences of people who lived and died centuries ago.

However, many ancient pieces of writing remain frustratingly closed books because their script is undeciphered. These include the Indus script, proto-Elamite from Iran, Linear A from the Minoan civilisation on Crete and Rongorongo from Easter Island. What is more, many spoken languages have never been written down. Today, although some 7,000 languages are spoken, only a fraction of these have a tradition of writing.

Lost and found

Some scripts were deliberately invented to fill the gap. The Korean script Hangul was created in the 1440s; the Cherokee script was invented in the 1820s. The most recent addition is Kodava, a language spoken (and now increasingly written) by 200,000 people in India. The script was invented in 2005.

To keep a writing system going requires a heavy investment, and the list of writing systems that have died out is a long one. Yet overall there is actually a huge upsurge in the use of writing as people increasingly use mobile phones and computers to communicate. Historians of the future will have access to much more of our thoughts, though whether they will be worth reading is another matter. OMG! RBTL!

THE LITERATE BRAIN

Reading is one of the most amazing things you do. Unlike spoken language, which human brains are pre-programmed to acquire, learning to read requires years of deliberate effort. The brain has to be taught to use cognitive modules that evolved for other reasons – pattern recognition, for example – to convert strings of abstract symbols into language. Writing is similarly hard and unintuitive. The time and effort required is why mass literacy is a very recent phenomenon.

HOW DID WE DISCOVER NOTHING?

A MAN USED TO OWN SEVEN GOATS. He bartered three for corn, gave one to each of his three daughters as dowry, and lost one somewhere. How many goats does he have left?

This is not a trick question. Oddly, though, for much of history, humans did not have the mathematical wherewithal to supply an answer. There is evidence of counting that stretches back five millennia. Yet even by the most generous definition, a mathematical conception of nothing – zero – has existed for less than half that time.

The story of zero is the story of counting and mathematics. But it is a tangled story of two different zeros: zero as a symbol to represent nothing, and zero as a number that can be used in calculations and has its own mathematical properties. It is natural to think the two are the same. They are not.

Zero the symbol was the first to appear. This is the zero familiar from a number such as 2016.

To know what 2016 means you have to grasp the concept of a 'positional number system'. Fortunately it isn't hard. Any primary school pupil who has mastered hundreds, tens and units has done it. The 6 in 2016 denotes six, the 1 means ten, and the 2 means two thousand. Zero's role is pivotal: it tells us that there are no 'hundreds' required in this number. Were it not for its presence, we might easily mistake 2016 for 216 or 2160.

The first positional number system was used to calculate the passage of seasons and years in Babylonia from around 1800 BC onwards. It used base 60, not the base 10 we are familiar with, so a hypothetical Babylonian primary school pupil would have to learn about 3600s, 60s and units. The system worked well but for one glaring flaw: if there was nothing to put in a column, the Babylonians just left a gap. That opened the door to numerical confusion.

Around 300 BC, presumably to stamp out such errors, the Babylonians introduced a new symbol, to denote an empty column. This was the world's first zero. Seven centuries later, on the other side of the world, it was invented a second time by Mayan priest-astronomers.

As a placeholder in a counting system, zero is clearly a useful concept. But neither the Babylonians nor the Mayans realised quite how useful it could be as a number in its own right.

Zero is admittedly not an entirely welcome addition to the pantheon of numbers. Accepting it invites all sorts of conceptual novelties that, if not handled with care, can bring the entire system crashing down. Unlike any other number, adding zero to something (or taking it away) doesn't change it. But multiply any number by zero and it collapses down to zero. And let's not even delve into what happens when we divide a number by zero.

Avoiding the void

Classical Greece, the next civilisation to experiment with the concept, was not keen. Greek thought was wedded to the idea that numbers expressed geometrical shapes; and what shape would correspond to something that wasn't there? Their world view also saw the planets and stars as embedded in a series of concentric celestial spheres, all centred on Earth and set in motion by an 'unmoved mover'. There was no place in this cosmology for a void, so it followed that zero was a godless concept. This was a picture later eagerly co-opted by Christian philosophy.

Eastern philosophy, rooted in ideas of eternal cycles of creation and destruction, had no such qualms. The next great staging post in zero's journey is found in the *Brāhmasphuṭasiddhānta*, a treatise on the relationship of mathematics to the physical world written in India around AD 628 by the astronomer Brahmagupta.

Brahmagupta was the first person who treated numbers as abstract quantities separate from physical or geometrical reality. This allowed him to consider new questions, such as what happens when you subtract from one number a number of greater size. In geometrical terms this is a nonsense: what area is left when a larger area is subtracted? But as soon as numbers became abstract entities, a new possibility opened up – the world of negative numbers.

From zero to hero

The result was the concept of a continuous number line stretching as far as you could see in both directions, with both positive and negative numbers. Sitting in the middle of this line, at the boundary between positive and negative, was sunya, the nothingness. This new number was soon unified with zero the symbol. It marked the birth of the purely abstract number system now used throughout the world, and soon spawned a new way of doing mathematics: algebra.

The news took a long time to reach Europe. It was only in 1202 that Italian mathematician Fibonacci presented details of the new counting system and demonstrated its superiority over the abacus for the performance of complex calculations. Merchants and bankers were quickly convinced, but the authorities were not. In 1299 the city of Florence banned numeral zero; they considered the ability to inflate a number's value by 10 simply by adding a zero on the end to be an open invitation to fraud.

It took the Copernican revolution – the crystal-sphere-shattering revelation that Earth moves around the Sun – to slowly shake European mathematics free of the shackles of Aristotelian cosmology from the sixteenth century onwards.

Thus a better understanding of zero became the fuse of the scientific revolution that followed. Subsequent events have confirmed just how essential zero is to mathematics and all that builds on it. Looking at zero sitting quietly in a number today it is equally hard to see how it could ever have caused so much confusion and distress. A case, most definitely, of much ado about nothing.

SIGNIFYING NOTHING

The concept of zero was invented in India 14 centuries ago, but it seems that the mathematicians of the time somehow managed without the symbol 0. In 662 BC Syrian scholar Severus Sebokht wrote that the great Indian mathematicians did calculations 'by means of nine signs' – presumably 1 to 9. The first record we have of a number zero appears 214 years later – a squashed-egg symbol recognisably close to our own, in an inscription at a temple in Gwalior, northern India.

WHEN DID WE GET THE MEASURE OF THINGS?

If you can't measure it, you can't manage it. That dictum runs through our society, from supermarkets to science. And it was ever thus: all ancient cultures seem to have invented systems to measure distance, weight, volume, area and time. History is littered with their now obsolete units: bushels, cubits, chains, roods, hundredweights and so on.

Measurement, unsurprisingly, is an exact science. Any system must be based on standard units that are accepted by everyone and easy to get hold of. That is why they were often based on the human body: a cubit is the distance from the elbow to the tip of the middle finger, for example. We still use units derived this way, including the foot and the hand.

Another approach was to use relatively uniform natural phenomena: gemstones were weighed in carob seeds, for example, which eventually became carats.

These were useful up to a point, but ultimately turned out to be too variable. So standards came to be written in stone or metal and stored in government buildings such as the Acropolis. There you would find precise definitions of units such as the dactyl (finger) or kochliarion (spoonful).

Revolutionary measures

The modern era of metrology was born in the storm of the French Revolution. Eager to establish a new national identity, its leaders wanted to sweep away all vestiges of the old system, including its irrational collection of weights and measures. The result was the neat, orderly metric system.

The original system consisted of just two units: the kilogram and the metre. The kilogram was defined as the mass of a litre of water at the temperature of melting ice, which instantly tied it to the metre (a litre is a cube with 10 cm sides).

The metre was defined as one ten-millionth of the distance from the North Pole to the equator. Unsurprisingly, establishing its actual length was no mean feat. It took seven years of climbing church towers from Dunkirk to Barcelona, triangulating the distance between them and observing the position of the pole star to work out the distance from pole to equator.

Both units were later enshrined in metal: a platinum cylinder weighing exactly a kilogram and a platinum bar exactly a metre long.

It was the most accurate and scientific system yet devised but was still dependent on variable quantities. Even before the metre was adopted as an international standard in 1875, there were grumbles that the measure was too vague. Physicist James Clerk Maxwell, for example, argued that units defined according to the dimensions of the Earth were inherently unstable because the surface was constantly changing.

In the 1870s US mathematician Charles Sanders Peirce had a key insight: the metre could be defined by the wavelength of light. This planted the idea of basing measurement on fundamental constants of nature. From that tiny seed grew today's scientific measuring system, the Système International d'Unités, commonly known as SI.

It took time, though. It was not until 1960 that the guardians of measurement finally acted. They shelved the platinum bar and replaced it with a metre based on the emission spectrum of an atom of krypton-86. This was in turn superseded in 1983 by defining a metre as the distance that light travels in 1/299,792,458 of a second.

The year 1960 also saw the birth of the SI system. As well as the metre, it defined an international standard for six other 'base units': kilograms, seconds, kelvins, amperes, moles and candelas (which measure light intensity). These base units can be combined to create all other units of measurement, such as joules, hertz, watts and ohms. There are 20 such named 'derived units'.

But that was not the end of metrology's problems. Five of the seven base units still had unsatisfactory definitions. The second, for example, was linked to the Earth's rotation, which varies slightly. That was fixed in 1967, but the other four were – and still are – problematic. If the SI system were a person, you would say they had a weight problem, an embarrassing mole, an unhealthy temperature and a distinct lack of spark.

That's a serious problem. Units of measurement need to be the same for everyone, everywhere, whether you're buying vegetables or doing particle physics. Not having a universal system can lead to disaster, such as when NASA lost its $125 million Mars Climate Orbiter in 1999 after a mix-up between imperial and metric units.

The weightiest problem is the kilogram, which is still defined by a physical object – a cylinder of platinum-iridium alloy cast in the 1870s.

Around 40 other kilograms were cast at the same time. Some are kept at the HQ of the International Bureau of Weights and Measures in Sèvres, near Paris, others at standards laboratories around the world. They are occasionally compared with one another. In 1949 metrologists found that the prototype and its companions had drifted apart by about 50 micrograms, an embarrassingly large error. They checked again in 1989 and the problem had not gone away.

Forever the same

The kilogram problem spills over to the mole, the unit that chemists use for measuring the amount of a substance. It is defined as the number of atoms in 0.012 kilograms of carbon 12. Doh!

The unit of temperature, the kelvin, is also not fit for purpose. It is defined as 1/273.16 of 0.01°C, the triple point of purified seawater at which ice, liquid and vapour can coexist. That's fine for most purposes but, for technical reasons, makes it hard to measure very high or very low temperatures. The ampere has even worse problems.

Metrologists are well aware of the problems and are busy tying all base units to unchanging constants of nature. When the revamp finally arrives, measurement will have rigorous foundations for the first time in history.

IMPOSSIBLE MEASUREMENT

The most imprecisely defined SI unit of all is the ampere, the unit of electric current: 'that constant current which, if maintained in two straight parallel conductors of infinite length, of negligible circular cross-section, and placed 1 metre apart in vacuum, would produce between these conductors a force equal to 2×10^{-7} newton per metre of length'. Did you get that? If not, never mind: it's an impossible measurement to make. Where do you get conductors of infinite length?

WHO STARTED KEEPING AN EYE ON THE TIME?

GLANCE AT THE SECOND HAND OF A CLOCK AND YOU MAY SEE SOMETHING STRANGE. Instead of marching onwards, it sometimes appears to freeze momentarily ... before tick-tocking back into action. This 'frozen time' illusion is caused by the design of the visual system: it shuts down when you move your eyes quickly, and the brain backfills what it guesses it must have missed. But it neatly encapsulates our problems with measuring time.

At the best of times, time is an elusive phenomenon. It flies, it drags, and sometimes it seems to stand still. We experience it as an inexorable conveyer belt of 'nows' stretching from the past into the future, but we are not sure it really exists: it may be a fundamental property of the universe, like space or mass. Or it may be an illusion created by our brains.

But this slipperiness has not stopped us from trying to nail it down. Humans being humans, we have invented ever more accurate ways to formalise and measure it.

Our ancient ancestors cannot have failed to notice the experience of now, and the predictable cycling of days, seasons and years. But for most of human prehistory natural timekeepers – dawn, dusk, the phases of the Moon and so on – were accurate enough. Megalithic monuments such as Stonehenge may be calendars of some kind, used to predict the passage of the seasons, but we can only speculate.

Clocking on

The first attempt we know of to create a formal system of timekeeping goes back around 4,000 years to when the ancient Egyptians came up with the idea of dividing the day into smaller units. Early sundials discovered in the Valley of the Kings tell us that daylight was divvied up into 12 equal portions, presumably to keep track of work time for labourers building the tombs. If so they, too, would have experienced time seeming to drag, especially in midsummer. The 'proto-hour' would have varied with day length, with a midsummer hour 16 modern minutes longer than a midwinter hour. Perhaps to solve this problem the ancient Egyptians also invented the water clock, which kept time independently of the Sun and divided a day into 24 equal portions.

The next big innovation was the creation of finer subdivisions. First out of the blocks were the Babylonians around 300 BC. They chopped the day up into three lots of 60, creating units corresponding to 24 minutes, 24 seconds and 0.4 seconds.

Seconds out

The system we use was invented around the end of the first millennium AD when Persian polymath Al-Biruni took the Egyptian concept of the 24-hour day and subdivided it by 60 twice, creating minutes and seconds – so-called because it is the second division by 60.

The second is still the fundamental unit of time. For centuries it retained its link with the solar cycle, defined as 1/86,400 of a day. But it gradually dawned on scientists that this definition revived the problem encountered by the ancient Egyptians. A day is not always precisely the same length. The gravitational pull of the Moon and the Sun are gradually slowing down Earth's spin, so 100 years ago the day was a little shorter and 100 years from now will be a little longer. Instabilities due to the atmosphere and Earth's roiling core also mean the planet's spin can slow down or speed up in unpredictable ways.

For everyday purposes this is not a problem. But with so many other units of measurement reliant on the second a variable second wasn't acceptable. Scientists eventually hit on a similar solution to the Egyptians, albeit one that was more sophisticated than dripping water.

In 1967, after years of often arcane debate, the International Committee for Weights and Measures agreed on a new definition of the second. Henceforth it would be the 'atomic second', defined by a specific number of vibrations of a caesium atom. This decision was a decisive break in the historical link between astronomy and time.

But astronomy still makes its unpredictable presence felt every few years. We can measure the passage of time with an atomic clock all we like, but the day continues to lengthen imperceptibly, leading atomic time and Earth time to gradually drift apart.

A leap into the unknown

From a human perspective the difference is negligible, two or three minutes per century. But for science this is way too inaccurate. So in 1972, the leap second system was born. Astronomers track Earth's rotation using the most stringently fixed reference point they can find – quasars that are billions of light years away. When a variation in Earth's spin threatens to send Earth time more than 0.9 seconds astray from atomic time, they issue a call to add or subtract a leap second. So far, the order has always been to add. The result is Coordinated Universal Time, or UTC.

Meanwhile, the drive to measure ever finer divisions of time continues. Atomic clocks are staggeringly accurate: the first reliable one, invented in 1955, was accurate to 1 second in 300 years.

That has since been surpassed many times. In 2013, US scientists built an atomic clock, which, had it been ticking since the Cambrian Explosion 542 million years ago, would have lost or gained only about half a second. The latest technology promises even more. A new generation of tickers, known as optical clocks, will soon be so refined that if one had clocked every second since the Big Bang 13.8 billion years ago it would still be bang on time.

DEEP TIME

One of the most conceptually difficult breakthroughs in our measurement of time was realising just how much of it there is. From the perspective of a lifetime, a millennium is just about conceivable. But 13.8 billion years is beyond comprehension. 'Deep time' is so counter to the grain of common sense that it took about 4,000 years from the invention of timekeeping itself to discover it.

Until the mid eighteenth century the universe was assumed to be a few thousand years old. Then it gradually dawned on geologists that this was off by several orders of magnitude. The rocks they studied appeared eternal and unchanging, but this was an illusion. In fact, the strata, the fossils and the fault lines spoke of unimaginably slow changes unfolding over mind-boggling stretches of time.

WHEN DID WE START ARGUING ABOUT POLITICS?

IF YOU HAVE EVER WATCHED RIVAL POLITICIANS SLUGGING IT OUT AND THOUGHT 'THEY SEEM TO LIVE IN DIFFERENT WORLDS', YOU'RE NOT FAR WRONG. Their disagreements go deeper than ideology: they're biological.

Humans are political animals. Despite its modern association with career politicians and government, politics is really no more than the perpetual debate over how to organise society and distribute power and resources. That argument has been raging for millennia. Nomadic bands of hunter-gatherers need to make those decisions just as much as we do.

In pre-modern societies politics largely consisted of power struggles between warlords. But as societies became more civilised, the struggles took on a more democratic character. And what tends to emerge is something that was first noted in France in the last decade of the eighteenth century. During those revolutionary times French society fractured along a clear faultline. One side supported the monarchy, the church and the other institutions of the *Ancien Régime*. The other supported the revolution. The traditionalists sat on the right side of the legislative assembly and the revolutionary faction on the left.

Power struggles

To a lesser or greater degree, all politics before and since mirrors this fundamental dividing line. Politics can be understood as a struggle between two competing impulses: to defend the status quo, or to overturn it. Try to think of a modern political system that isn't defined by a struggle between right and left, conservatives and progressives. Where does this seemingly universal human divide come from?

The conventional wisdom is that political leaning is something we consciously and rationally choose based on evidence and argument. If we differ, it is because we reason to different conclusions. Recent research, however, suggests that this is far from the whole story. Politics is in the blood, and political differences are deeply rooted in basic biology. Not only that, they are largely beyond conscious control.

Research on the biological roots of political persuasion first appeared in the 1950s as the world struggled to understand totalitarianism. It is mainly remembered for identifying something called the authoritarian personality. The idea that this applied to any more than a tiny fraction of the population was disputed, however, and interest waned.

But the researchers were on to something. Modern studies have found that personality does influence political belief. When psychologists snooped around offices and dorm rooms, they found that conservatives and progressives tend to organise their spaces differently. Conservatives favour tidiness and conventionality and have more objects related to order. Progressives' rooms are more cluttered and have more objects related to exploration.

Moral maze

The researchers concluded that these outward differences were a manifestation of inward personality traits – openness to experience and conscientiousness, two of the 'big five' dimensions that are known to have a strong genetic basis.

Several related studies have shown that conservatives have a higher need for 'cognitive closure' – wanting to turn uncertainties into certainties and ambiguity into clarity.

Another area where biological differences have been found is in moral judgements. Progressives find suffering and inequality morally offensive, while conservatives are more bothered by disrespect for authority and tradition and signs of sexual or spiritual 'impurity'. Again, these differences have surprising biological roots: they have been linked to how easily people are disgusted.

As a rule, conservatives are more easily disgusted by stimuli like fart smells. Disgust tends to make people of all political persuasions more unforgiving of morally suspect behaviour, but conservatives react more harshly. This may explain differences of opinion over gay marriage and illegal immigration. Conservatives often feel strong revulsion at these violations of the status quo and so judge them to be morally unacceptable. Liberals are less easily disgusted and less likely to judge them so harshly.

Differences have even been found in the way people see the world, for example in their reaction to frights. Conservatives have a more pronounced startle reflex to sudden loud noises, blinking harder and sweating more. They also show stronger responses to threatening images and look at them more rapidly and for longer. Conservatives are more likely to report seeing the world as a dangerous place.

Most controversially, scientists have begun to search for the genetic roots of these differences. For 25 years we have known about the high heritability of political attitudes. Identical twins are much more likely to share political views than fraternal twins, suggesting it is not only their shared environment that is at work but also their shared genes.

Right, left and centre

More recently, geneticists have begun looking at particular genes that may contribute to ideology. Nobody is suggesting that there are genes 'for' progressivism or conservatism, but one gene of interest is the 7R variant of the *DRD4* dopamine receptor gene, which has been associated with novelty-seeking behaviour and left-leaning politics.

The research has been criticised for reducing the nuances of political opinion to a simplistic binary and treating conservatism like a personality disorder. The real world is more complex, with views arrayed on a spectrum and numerous strands of opinion within the broad churches of left and right. And there are other political traditions that don't fit the model, notably libertarianism.

Nonetheless, there is substantial evidence that politics is driven not by differences in opinion so much as by differences in basic biology. So instead of being angry with your political opponents, you ought to feel sorry for them: they really can't help being so totally, utterly, stupidly wrong.

THE COMPANY WE KEEP

Some of the clearest evidence that our political beliefs are not conscious decisions comes from psychological tests that measure unconscious attitudes – that is, preferences that operate outside of awareness. They show that people of different ideologies also differ in their social preferences. As a rule, conservatives are more likely than progressives to prefer high-status people and dominant social groups such as whites and heterosexuals. Progressives are more comfortable than conservatives with members of ethnic and sexual minorities – though it is important to note that progressives also unconsciously prefer high-status groups, just not as much as conservatives do.

WHEN DID ALCHEMY BECOME SCIENCE?

THE CHEESEMAKERS WERE PROBABLY NONE TOO IMPRESSED, BUT THE REST OF THE WORLD SHOULD BE ETERNALLY GRATEFUL.

It was 17 February 1869 and Russian chemist Dmitri Mendeleev was due to do some consultancy work at a cheese factory in St Petersburg. But he cancelled and spent the day scribbling feverishly at home. By the evening he had the outline of one of the most successful scientific theories of all time: the periodic table of the elements.

Mendeleev's eureka moment was the culmination of centuries of work trying to understand and control the processes of material change. What happens when a candle burns? Why does a pinch of salt disappear when stirred into a glass of water? Can lead be turned into gold? We now recognise those questions as belonging to the realm of chemistry, which has a reputation as a rather dull and sober science. But its origin was anything but.

The first steps were made by philosophers in ancient Greece. Aristotle asserted that everything was made from four elements: earth, fire, air and water. Materials had particular qualities because of the proportion of these elements that they contained. A metal, for instance, was made from earth and water, but if you heated it some of the earth changed to fire.

Metals and dyes

Aristotle died in 322 BC, a decade after Alexander the Great conquered Egypt and established a new capital, Alexandria. Artisans steeped in Aristotelian philosophy began to dabble in metallurgy and dye-making. They called their craft khymeia, meaning 'cast together'. The tradition was later passed to Islamic scholars, who called it al-khimya. Their knowledge eventually found its way to medieval Europe, where practitioners of magick wrapped it in mysticism and called it alchemy or just chymie.

The alchemists' main goal was the philosopher's stone, a substance which could transmute base metals into gold and silver, cure any disease and held the key to eternal life. They were also craftsmen who used their expertise at manipulating and transforming materials to produce medicines, glass and explosives.

But alchemy wasn't a science. That turning point came in 1661 when philosopher Robert Boyle published a groundbreaking book called *The Sceptical Chymist*, which applied the newly-minted scientific methods to chymie. Boyle argued that you couldn't just assert that matter was made of four elements; you had to prove it with repeatable experiments.

Taming the elements

The man who provided them was French aristocrat Antoine Lavoisier. He took on Boyle's challenge and went searching for elements, which he defined as anything that could not be broken down further. In 1789 Lavoisier published a list of 33 'elements', many of which actually *are* elements as we understand the concept today. Many more were soon discovered. The idea that each element had its own unique atom became popular, as did the idea that the elements combined with each other to form compounds.

By Mendeleev's time, 63 elements were known. His breakthrough was to organise them into groups by atomic weight and thereby reveal some patterns in their properties. Group 1, for example, were all soft metals that react violently with water. Group 7 included the gases fluorine, chlorine and bromine, which exist as molecules composed of two atoms. That was not the only pattern. Within each group the reactivity of the elements changed as the atoms got heavier. In group 1, reactivity increases as the atoms get heavier. But in group 7 the opposite is true.

The periodic table is the unifying theory of chemistry. It didn't just explain observations, it also made predictions. Where there was no element with the correct properties, Mendeleev boldly left a gap, claiming that a new element would be discovered to fill it. He was right. For example, there was a space immediately below silicon. Mendeleev called it 'eka-silicon' and 15 years later German chemist Clemens Winkler discovered it. He called it germanium.

Modern alchemy

It was not long until the underlying cause of the patterns was discovered: the electron. The particle was discovered in 1896 but the crucial experiments were done by Hans Geiger and Ernest Marsden a decade later. They fired a stream of helium nuclei at a piece of gold foil. Unexpectedly, many of the nuclei passed straight through, leading to the conclusion that gold atoms were mostly empty space. Their interpretation, which later proved essentially correct, was that the electrons were orbiting the nucleus, leaving huge tracts of nothingness in between.

The orbits of the electrons explain an element's chemical and physical properties. Reactivity, for example, depends on how easily an atom can gain or lose an electron.

But the electron is something of a double-edged sword. For all the regularity it introduced, the shorthand chemists use to understand its effects is just an approximation. In truth, electrons are quantum objects with weird properties: they can be in two places at once, or 'tunnel' through space.

As the quantum revolution gathered pace, scientists also started to probe the atomic nucleus in detail. One of their key discoveries was that elements could be 'transmuted' from one to another by nuclear reactions – something that appeared forbidden by the laws of chemistry. Nobody mentioned the A word, but in 1951 chemist Glenn Seaborg took a base metal, bismuth, and turned it into gold.

NEWTONIAN MAGIC

The medieval alchemists of Europe included not just mystics but also respectable scholars. The most respectable of all was Isaac Newton. In the 1680s he wrote a dictionary of alchemical terms called *Index Chemicus.*

It seems odd to think that someone we regard as a great scientist would have been seduced by mumbo jumbo, but at the time there was no clear distinction between science and magick. Nonetheless, by the time Newton died, in 1727, alchemy had become a dark art. Newton left a huge library of unpublished notes and papers, many on alchemy. When they were posthumously examined by Thomas Pellet, a member of the Royal Society, he decided to suppress them, tagging them with the instruction 'Not fit to be printed'.

HOW DID WE DISCOVER THAT REALITY IS SO STRANGE?

In 1874, when a 17-year-old science prodigy called Max Planck told his university professor he wanted to embark upon a career in physics, the older man proffered some advice. 'In this field, almost everything is already discovered,' he pronounced, 'and all that remains is to fill a few holes.'

It turned out he was sort of right – though the holes were rabbit holes that even Lewis Carroll would have thought odd. Within a few years, attempts to fill them had given rise to a revolutionary but mind-bending new understanding of the universe. And the man who went down them first, albeit reluctantly, was Planck.

Today quantum mechanics is our most successful description of reality. It allows us to understand everything from atoms to stars. It has also taught us that reality is fundamentally and deeply mysterious, maybe even incomprehensible.

Light bulb moment

The revolution started, appropriately, with a light bulb. In 1894, Planck – by then a university professor in Berlin – was commissioned to do some technical work on Thomas Edison's new invention. Electric companies wanted to know how to squeeze the maximum amount of white light from the minimum amount of energy, and so Planck started to explore the relationship between the temperature of the filament and the colour of the light.

This turned out to be a restatement of a known head-scratcher called the black body radiation problem, which described the relationship between the temperature of an object such as a piece of metal and the colour of light it emitted (a black body is a theoretical entity that is both a perfect absorber and an emitter of electromagnetic radiation). Experimental measurements had uncovered a huge anomaly that physics could not solve: no matter how hot they became, black bodies emitted almost no ultraviolet light. This became known as the ultraviolet catastrophe.

In December 1900 the then 42-year-old Planck stood in front of the German Physical Society and proposed a solution: energy, rather than being a continuous phenomenon that can exist in quantities of any size, exists only in discrete lumps. Each one of these units he called a quantum. At the time Planck did not realise that he was crawling into a hole from which there was no escape. But his conclusion – which he described as an 'act of despair' – inspired a younger generation of physicists who were eager to go in deep.

One of them was a 25-year-old nobody called Albert Einstein. He was trying to understand the photoelectric effect, the phenomenon by which many metals spit out electrons when bathed in certain frequencies of light, regardless of its intensity. Planck's quanta were exactly the concept he needed. Einstein realised that the effect could only be explained if light, too, was quantised. If so, it was no longer possible to think of light in classical terms, as a wave propagating through space. Instead, it must consist of a stream of particles, each of which carried a single quantum of energy.

Neither one thing nor the other

This was a difficult idea for physicists to swallow, as there was unequivocal evidence that light was a wave. In particular, shining a light through two slits produced an interference pattern exactly like two sets of ripples on a pond. The only way to square the circle was to throw common-sense notions out of the window and accept the idea that light was both a wave and a particle.

By the 1920s it was clear that wave–particle duality was everywhere. This had physicists of the old school quietly seething. And there was worse to come.

In 1927 German theorist Werner Heisenberg realised that the consequences of wave–particle duality imposed a fundamental limit on how much information we could ever know about the world. The more precisely we measure a particle's location, for example, the less we can know about its momentum. In the quantum world particles are not like snooker balls; they do not possess two separate properties called location and momentum, but a mixture of the two that cannot be teased apart.

Heisenberg's uncertainty principle remains one of quantum theory's most counterintuitive predictions. And as his ideas developed, they became ever more unhinged from everyday reality.

Theatre of the absurd

Many found the work of his Austrian rival Erwin Schrödinger more palatable. He agreed it was impossible to describe a particle as inhabiting a fixed point in space. Instead the best we could hope for was to assign a set of probabilities to all the possible positions where it was likely to exist. By this logic a particle would only settle into a specific location once somebody took the trouble of looking at it.

This concept of a superposition of states that only collapse under observation became a key tenet of the Copenhagen Interpretation of quantum mechanics, formulated by Heisenberg and Niels Bohr. It also led to another important but totally bizarre concept – entanglement, the superposition of two particles at great distance from one another.

Such absurdities weighed heavily on the quantum pioneers. As Bohr himself put it, 'those who are not shocked when they first come across quantum theory cannot possibly have understood it'.

But they were vindicated in style. A new era of experiments verified even the most mind-boggling predictions. But though understanding has increased in leaps and bounds, our bewilderment is just as great as it has ever been. There are holes all right, but they may never be filled.

CAT IN THE BOX

In the 1920s the Copenhagen Interpretation of quantum mechanics became the most popular way of coming to terms with the weirdness of the quantum world, but not everybody was comfortable with its implications. Erwin Schrödinger called attention to its absurdity with a famous but much misunderstood thought experiment. Imagine a cat locked in a box with a vial of poison that has a 50 per cent chance of shattering. According to quantum mechanics, until the box is opened both outcomes are equally possible. Until somebody peeks inside, the cat is both alive and dead at the same time.

PART SIX

INVENTIONS

WHY DID IT TAKE SO LONG TO INVENT THE WHEEL?

URUK, 5,500 YEARS AGO. The Sumerian city is a splendid sight, the largest and richest human settlement the world has ever seen. And it is distinctively urban, with tens of thousands of inhabitants, big buildings, city walls, markets and outlying residential districts.

That much we know from the city's ruins. But one of the essential trappings of a modern city is conspicuous by its absence: the wheel.

It is hard to imagine a functioning modern city without cars, taxis, buses, trucks, bikes and tuk-tuks to ferry goods and people about. But Uruk doesn't seem to be a city on wheels. The only evidence we have that it was are a handful of etchings that vaguely resemble four-wheeled wagons, inscribed on poorly dated clay tablets. In contrast, there are many pictographs of contraptions that look like sledges, suggesting that Uruk was a city driven by vehicles designed to be dragged along the ground.

If wheels really were rare or even absent in Sumeria, then that would be a bit of a head-scratcher. The technology seems so blindingly obvious, so easy to make, so obviously useful, and so ripe for invention. Potter's wheels were an ancient technology by this time. The city had streets that were flat and smooth enough to drag a sledge along and would have been ideal for wheels. Asses, oxen and other beasts of burden that could pull wagons had been domesticated, and complex trade networks had sprung up across the region. Metal-working was becoming commonplace. It wasn't the Stone Age, for heaven's sake!

Perhaps we simply haven't found the remains of Uruk's carts and wagons yet. Early vehicles were probably made of wood and rope, which do not preserve well in the archaeological record. Nonetheless, Sumeria's earliest unambiguous depictions of wheeled vehicles date from a thousand years later, on a decorated wooden box depicting four-wheeled chariots drawn by asses. The obvious conclusion is that whatever was making Uruk's world go round, it wasn't wheels.

Even more oddly, at the time the sophisticated Sumerians were apparently struggling to invent the bleeding obvious, people in relative backwaters were travelling. Under a 5,500-year-old tomb in Flintbek, Germany, archaeologists found a pair of parallel, wavy ruts that must be the tracks of a wonky-wheeled cart. The same culture – the funnel beaker people – made pots decorated with motifs that look very much like four-wheeled wagons.

The earliest physical remains of an actual wheel also come from backwards Europe. Discovered in a bog in what is now Slovenia in 2002, the Ljubljana Marshes wheel is a wooden wheel/axle combination dated to about 5,150 years ago. What it was attached to is not known; possibly a hand cart. Further east, on the steppes of what is now Ukraine, wheels and complete carts have been found in 5,000-year-old graves.

Talking about a revolution

We don't know whether Europe's ancient wagon technology was wheeled in from elsewhere or invented independently. But there is another line of evidence that the wheel was established in Europe very early on. Languages, like bones and DNA, contain traces of the distant past. In the same way that biologists can reconstruct the common ancestor of two species by looking at the genes and physical traits that they share, so linguists reconstruct extinct languages. The word 'name', for instance, derives from the Latin 'nomen', which also gives us the French 'nom' and the Spanish 'nombre'.

Creating a family tree of modern European languages shows that most of them, and some non-European languages, share a common origin: a now extinct language dubbed Proto-Indo-European. It probably arose somewhere in western Asia and was brought to Europe by an incoming population.

The original vocabulary reconstructed by linguists contains five words relating to the wheel. Two literally mean 'wheel', one means 'axle', one refers to a pole used to yoke animals to a cart and one is a verb for the action of transporting in a vehicle. Talking about the wheel so much implies it was an important part of the lives of its speakers.

Proto-Indo-European has been dated to about 5,500 years ago, suggesting that the wheel was already an old technology when the Ljubljana wheel was made. What is more, one group of the speakers of the language were the Yamnaya, the people who built the wagon graves in Ukraine.

Genetic evidence suggests that around 4,500 years ago the Yamnaya expanded west into central Europe and went on to found some of the continent's dominant late Neolithic and Copper Age cultures, including the vast Corded Ware culture – named after its distinctive pottery – that stretched from the North Sea to central Russia.

On the wagon

The Yamnaya were cattle herders. Up until about 5,500 years ago, their settlements clung to the river valleys of their homeland – the only place where they had easy access to the water they and their livestock needed. But their arrival in Europe is consistent with the idea that they had mastered wagon technology. With wagons they could take water and food wherever they wanted, and the archaeological record shows they began to occupy vast territories.

The wheel quickly spread. By about 4,500 years ago light and nippy two-wheeled chariots made their first appearance and were soon being used in warfare. The wheel also inspired more peaceful inventions such as water wheels, cogs and spinning wheels. It took a while, but once wheels were invented, technological civilisation was on a roll.

WHY DON'T ANIMALS HAVE WHEELS?

Evolution has come up with all kinds of elegant solutions to the problem of locomotion: birds fly, squid have jet propulsion, geckos scale walls and fleas have spring-loaded legs. And yet no animal has ever evolved wheels. Why not? The reason is that evolution works incrementally and doesn't have foresight: it only comes up with designs that are useful right here and right now. Each tiny step towards flight or jet propulsion was better than what went before. But there is no incremental step towards a wheel that is useful in itself. Not to mention no way to create an appendage that can rotate freely while still being supplied with blood vessels and nerves.

WHEN DID WE START TALKING OVER THE AIRWAVES?

PONTECCHIO, ITALY, DECEMBER 1895. A young Italian aristocrat awakens his mother in the early hours of the morning to show off his new creation. On a device he has secretly built in the attic of their villa near Bologna, Guglielmo Marconi taps out a message in Morse code. At the other end of the room, a bell rings out the message. Wireless contact has been made.

To modern eyes familiar with television, mobile phones and Wi-Fi, that might seem a rather feeble achievement. But Marconi's invention of a machine that could transmit signals over the airwaves rather than through a cable was one of the most influential technological breakthroughs of the twentieth century. So influential, in fact, that many other inventors tried to claim it as their own – often with some justification.

Received and understood

The key to Marconi's success was to combine two existing inventions to create a new one. The first was the transmitter, based on lab equipment used by German physicist Heinrich Hertz to demonstrate that it was possible to create electromagnetic waves. The other was the coherer, a receiving device invented by French physicist Édouard Branly to detect ambient electromagnetism such as that created by lightning.

After his attic experiment Marconi was a man on a (trans) mission. He was soon sending signals over long distances outdoors. In 1896 he moved to London and applied for a patent on his invention; a year later he founded the Wireless Telegraph & Signal Company which established the first international radio links and laid the groundwork for commercial radio broadcasting. In 1909 he was awarded a share of the Nobel Prize in recognition of his 'contributions to the development of wireless telegraphy'.

Marconi is widely recognised as the inventor of radio, but he was really just riding on a wave of innovation for which he somehow ended up scooping the credit. He was neither the first engineer to twig that wireless transmission was possible, nor the only one working on it. With a few twists of fate, the history books would tell a different story.

One scientist who definitely deserves more recognition is the German physicist Karl Ferdinand Braun, who shared the Nobel Prize with Marconi even though the two did not work together. Braun invented many of the technologies that Marconi would later rely on, and Marconi himself admitted to 'borrowing' some of his ideas.

Another serious rival was the flamboyant genius Nikola Tesla. In 1893, two years before Marconi's attic demo, Tesla gave a widely reported lecture at the Franklin Institute in Philadelphia describing in theory how to build a wireless transmitter and receiver. But at that stage he didn't have any equipment. It was, he said, 'a serious problem in electrical engineering which must be carried out some day'. Tesla tried to do it himself and eventually filed a patent in 1897, but Marconi had already beaten him to the prize.

Having one bona fide genius breathing down your neck is one thing; having two is quite another. Around the same time Marconi was tinkering in his attic, the brilliant New Zealander Ernest Rutherford was powering ahead at Canterbury College in Christchurch. But Marconi had a stroke of luck: in 1895 Rutherford moved to Cambridge to continue his work but was stymied when his laboratory suddenly decided to focus all its efforts on newly discovered X-rays.

Ahead of his time

Tesla and Rutherford went down in history for other reasons, but Marconi's other rivals have largely been forgotten. One was the English physicist Oliver Lodge, who perhaps has the most credible claim to have beaten Marconi. In August 1894 Lodge wirelessly transmitted some Morse code from Oxford's Clarendon Laboratory to the Oxford museum about 60 metres away. His hardware was remarkably similar to the one Marconi 'invented', though the Italian denied any knowledge of it.

Lodge was probably a victim of his own modesty: he called his work 'a very infantile form of radio telegraphy' and did not attempt to patent his ideas until 1897, by which time Marconi had the intellectual property sewn up.

Perhaps the bolshiest counterclaim to credit came eight years after Marconi's death. On 7 May 1945 a distinguished audience gathered at the Bolshoi Theatre in Moscow to hear that henceforth the date would be celebrated as 'Radio Day' in honour of Russian physicist Aleksandr Popov of the Naval Engineering College near St Petersburg. Fifty years earlier, the audience were told, Popov had achieved the first ever wireless transmission at a meeting of the Russian Physico-Chemical Society.

Propaganda machine

According to an account of the event by Soviet scientist Victor Gabel, published in *Wireless World* in 1925, Popov had wirelessly transmitted the words 'Heinrich Hertz' in Morse. The transmission predated Marconi's patent, making Popov the official inventor of radio.

If it actually happened, that is. Gabel's is the only account of the meeting; the journal editor was sceptical but published it anyway. Popov himself never claimed priority over Marconi, and does not even seem to have recognised him as a rival. The two men met in 1902 and became firm friends.

But Popov's modesty was more than compensated for by the bravado of the Soviet state – possibly spurred on by Marconi's membership of Italy's Fascist party. When the *Wireless World* article was published in 1925, the USSR's propaganda machine sprang into action. At that point Soviet science and technology was lagging badly behind the west, a fact that Stalin wanted to conceal from the public. Not only was Popov credited with the invention of radio, but Russian scientists were even said to have invented television and aeroplanes. The propaganda worked: a Russian textbook called *Fundamentals of Radio* published in 1963 does not even mention Marconi.

Whoever deserves credit, the invention of wireless telegraphy arguably created the modern world. TV broadcasts began in 1928, radar helped win the Second World War, and today's quintessential technology, the smartphone, began life as a two-way radio.

ARE WE ON AIR?

It did not take long for entrepreneurs to realise that Marconi's invention for wirelessly transmitting Morse code had potential as a mass-market technology. On 2 November 1920 the world's first commercial radio station, KDKA of East Pittsburgh, Pennsylvania, went on air. It broadcast the results of that day's presidential elections, and then put out a plaintive cry for customer feedback: 'We'd appreciate it if anyone hearing this broadcast would communicate with us, as we are very anxious to know how far the broadcast is reaching and how it is being received.'

WHO WAS THE FIRST PERSON TO FLY?

IF YOU HAPPEN TO BE PASSING THROUGH CHARD, A SMALL TOWN IN SOMERSET, ENGLAND, YOU MAY BE SURPRISED TO SEE SIGNS WELCOMING YOU TO THE 'BIRTHPLACE OF POWERED FLIGHT'. If you don't believe your eyes, head to the town centre. On the high street you can see a bronze statue commemorating the world's first aeroplane.

Every town needs a claim to fame, but doesn't that one belong to Kitty Hawk in North Carolina, where the Wright Brothers finally fulfilled humanity's centuries-old dream of flying like a bird?

Yes and no. Kitty Hawk certainly deserves its place in aviation history, but so does Chard. In June 1848, inventor John Stringfellow achieved the seemingly impossible when his steam-powered aeroplane flew the length of a disused lace mill in the centre of town.

Stringfellow thus came within a whisker of immortality, except for one thing: he himself did not fly. His aeroplane was what we would call a drone. It wasn't until Orville Wright achieved his 12-second, 37-metre flight in Kitty Hawk in 1903 that humans finally emulated the birds and achieved heavier-than-air powered flight.

The history of flight is full of near misses and near forgotten pioneers. But as Orville and Wilbur readily acknowledged, all of them paved the way for their eventual success. One of the most influential was the English gentleman scientist George Cayley, who would arguably have gone one better than his contemporary Stringfellow and achieved powered human flight a full 50 years before Wilbur and Orville – if only the engine technology had been up to the job.

In Cayley's youth, scientists and the public believed that it was not only impossible to fly like a bird, but a folly to even try. This did not discourage Cayley, even though his contemporaries thought he was bonkers. In 1799 he published a design for an aeroplane and also the earliest description of the aerodynamic forces on a wing that would enable it to fly. His three-part treatise *Aerial Navigation*, published in 1809 and 1810, was greeted with scepticism.

Cayley didn't care. He had completed a series of experiments to back up his calculations and was convinced that he had cracked the problem of powered flight. Cayley constructed increasingly sophisticated flying machines, culminating in a full-scale glider which his grandson, George, flew across a shallow valley near Scarborough in Yorkshire in 1853.

She flies like a bird

The craft had fixed wings and a rudimentary tail, plus a rudder at the rear to steer with. Cayley had realised that a bird's tail is crucial to its ability to fly, and that it would therefore also be essential for a flying machine. What it lacked was an engine – a device which he had spent many fruitless years trying to develop. Always the glass-half-full type, he opted for a glider instead.

The Wright brothers also cited two other pioneers as important influences on their achievement. One was Otto Lilienthal in Germany, who made gliders with wings that were highly curved on their top surface, like a bird's. These wings achieved levels of lift that no other experimenter had dreamed of. Lilienthal made numerous glider flights from various jumping-off places, including a specially constructed 15-metre hill near his home on the outskirts of Berlin. But he paid the ultimate price for his experiments, fatally breaking his neck when one of his gliders stalled and crashed in 1896.

The other was American astronomer Samuel Langley. In 1896 he made a model aircraft that was powered by a small steam engine. It flew for more than a kilometre before running out of fuel. But Langley was never able to make an aircraft big enough to carry a human pilot because full-size steam engines were just too heavy. It took the development of motor car technology to solve this problem for would-be aviators in the shape of the petrol-fuelled internal combustion engine.

By October 1903 Langley had given up on the steam engine and was trying to launch a petrol-powered plane from the roof of a houseboat in the Potomac River in Washington DC. His attempts failed miserably, mainly because he had not paid enough attention to the need to control the aircraft once it was airborne. Soon afterwards, the Wright brothers' petrol-engine aircraft lifted off. Unlike Langley, the Wrights had done their homework on control, learning from the experiences of others – and, of course, birds.

One of the smartest of the Wrights' developments was to work out a way of controlling the roll of the aircraft – its motion about an end-to-end axis through the fuselage. Instead of making the pilot shift his weight from side to side as Lilienthal had done in his fatally flawed glider, the Wrights designed their wings to warp under the command of the pilot, momentarily creating more lift on one side of the plane or the other.

Fly me to the Moon

This was an important step because the ability to steer the aircraft without a catastrophic loss of lift had been a major problem. In addition, the Flyer had a movable front-mounted aerofoil called an elevator, which controlled pitching – the up-and-down movement of the nose – and a rear rudder to control yawing, or side-to-side movement. Together, these gave the Wrights' aeroplane control of its movement in three dimensions, a crucial factor that none of their influential predecessors had properly accounted for.

Orville's brief flight remains one of the greatest achievements in technological history. Just 66 years later, NASA landed humans on the Moon and passengers were routinely boarding airliners to fly to the other side of the world. Flying is now so routine, so laborious and so unglamorous that it is easy to forget how much our ancestors yearned to do it. To them, a cattle-class flight on a budget airline would have been little short of a miracle.

WE HAVE DRIFT OFF!

Humans have always yearned to fly, and in October 1783, they finally did. We're not entirely sure who went first – it was either Jacques-Étienne Montgolfier or his collaborator, Pilâtre de Rozier. But whichever one it was, he climbed into the basket of a hot air balloon near Paris and went up into the sky.

It was a sensational achievement, but also a bit of an anticlimax. Yes, humans had always yearned to fly – but *like a bird*. Bobbing about in a hot air balloon – a lighter-than-air craft without an engine – did not really count. A month later de Rozier made the first untethered balloon flight, drifting gently for 25 minutes above Paris. In 1785 he achieved another aviation first: he became the first person to die in an air crash when his balloon burst and fell out of the sky near Calais.

WHY ARE WE STUCK WITH THE QWERTY KEYBOARD?

TECHNOLOGY OFTEN CONTRIBUTES NEW WORDS TO THE ENGLISH LANGUAGE: TELEVISION, HOOVER AND IPOD TO NAME A FEW. But none have origins quite like the word QWERTY.

As one of the world's most ubiquitous technologies, used by billions of people every day, we rarely give a second thought to computer keyboards. But behind the ordinariness and familiarity there's something very odd about them. Why are the letters arranged that way?

The world's love–hate relationship with the QWERTY keyboard began in a small workshop in Milwaukee in 1866. That was where a publisher called Christopher Latham Sholes began work on an invention he hoped would make him rich: a machine to automatically number the pages of books.

Sholes was joined by an inventor friend called Carlos Glidden. In July 1867, he happened to read a short description of a 'type-writing machine' in *Scientific American*. It appears to have inspired them to change course and create 'a machine by which … a man may print his thoughts twice as fast as he can write them'.

Piano or typewriter?

A year later they were in possession of three patents. You'd be hard pressed to recognise their creation as a typewriter, though. It looked more like a piano, with ivory and ebony keys, one for each letter.

The machine was prone to jamming and the lines of type tended to drift off course, but Sholes used it to write to potential investors. One of them, James Densmore, immediately bought a quarter share of the patents, sight unseen. But when he got to Milwaukee to cast an eye over his investment, he was less than impressed, declaring it to be 'useless'. Nonetheless, Densmore believed in the general idea and urged Sholes to continue.

What happened next is a little murky. Sholes filed another patent in 1872 which shows the piano keyboard had been dropped in favour of rows of circular keys, but it did not specify which letter was where.

Then, almost out of the blue, QWERTY (almost) appeared. In August 1872 *Scientific American* published a glowing article about the 'Sholes' Type Writer', illustrated with an engraving of the machine showing a four-row keyboard with a second row starting QWE.TY.

QWERTY comes together

Densmore demonstrated the typewriter to engineers at E. Remington & Sons, a gunmaker based in New York that had branched out into home appliances. Remington signed a contract to manufacture the machine, and produced a prototype with another slightly different keyboard: QWERTUIOPY.

Sholes was apparently unhappy and demanded that the Y be reinstated between the T and the U. Remington agreed, and for the first time QWERTY came together. Remington put its No. 1 Type Writer on to the market in 1874 and it quickly became the world's first commercially successful writing machine.

By 1890 there were more than 100,000 QWERTY keyboards in use in the US. QWERTY, then, clearly evolved gradually from an initial design in the 1870s. But where did the arrangement come from?

One often-repeated explanation is that it was designed to 'slow the typist down' in order to stop the mechanism from jamming, a bug that dogged earlier designs. This was supposedly achieved by keeping common letter pairs apart.

But that cannot be true. E and R, the second most common letter pair in English, are next to one another. T and H, the most common of all, are near neighbours. A statistical analysis in 1949 found that a QWERTY keyboard actually has more close pairs than a keyboard arranged at random.

Another urban myth is that it enabled salesmen to impress customers by rapidly typing 'TYPE WRITER QUOTE' from the top row. It's a nice idea – and it does seem unlikely that these letters would appear together by chance – but there is no historical evidence for it.

Perhaps a more convincing though prosaic reason is that the keyboard is simply a semi-random rearrangement of the original piano-style keyboard.

We'll probably never know. A century after Sholes finalised

the keyboard, historian Jan Noyes of Loughborough University published a lengthy analysis concluding: 'There appears … to be no obvious reason for the placement of letters in the QWERTY layout.'

Worst possible design

One thing is clear: the keyboard was not designed with touch-typists in mind. As Noyes pointed out: 'the original QWERTY keyboard was intended for "hunt and peck" operation and not touch-typing'. Touch-typing was a later invention.

That may explain the well-known practical shortcomings of QWERTY. In the 1930s, as typewriters and typing became more common, researchers began questioning its usefulness. One fierce critic, educational psychologist August Dvorak (a distant cousin of the composer), had a team of engineers test 250 keyboard variations and concluded that the QWERTY design was one of the worst possible arrangements.

He had ulterior motives. In 1936 Dvorak had patented an alternative, the Dvorak Simplified Keyboard. He claimed it was easier to master, faster to use and put less strain on the hands. But it did not take off. In fact, since becoming the de facto standard, the QWERTY keyboard has seen off dozens if not hundreds of supposedly superior competitors. It transferred seamlessly from mechanical typewriters to computers and now on to touch screens, and is ubiquitous wherever the Latin alphabet is standard.

Yet despite its dismal reputation, it is not all that bad: a study in 1975 found that an accomplished typist could attain over 90 per cent of the theoretical maximum speed.

The real reason for its stubborn persistence is inertia: imagine the cost of designing, testing and manufacturing an alternative – and then retraining billions of people to use it. As long as keying letters into a machine is needed, QWERTY is here to stay.

FASTEST FINGERS IN THE WEST

The inventors of the typewriter set out to create 'a machine by which … a man may print his thoughts twice as fast as he can write them'. By one measure they succeeded. Handwriting rarely exceeds 30 words per minute and even poor typists can beat that. But by another they failed. Even the fastest typists cannot outpace shorthand.

HOW DID WE HARNESS ELECTRONICS TO DO MATHS?

IF YOU SAID THE WORD 'COMPUTER' TO SOMEBODY 70 YEARS AGO, THEY WOULD HAVE THOUGHT NOT OF A MACHINE ON A DESK BUT OF A PERSON BEHIND ONE, PENCIL AND PAPER IN HAND. In those days computers were people – usually women – who performed laborious calculations to meet the world's demand for pre-crunched numbers.

The fruits of their labours were books of mathematical tables that were an indispensable tool of the age. Whenever a scientist, engineer, navigator, banker or actuary needed to perform a complicated calculation, they reached for their tables.

These were perhaps the most tedious books ever published – a telephone directory would have been a riveting read in comparison – but that didn't bother a mathematician called Charles Babbage. He was a man of independent means who had his finger in many scientific pies. Among other things, he was a collector of printed mathematical tables and a ruthless ferreter-out of errors within them.

Useful as they were, the tables were riddled with mistakes. A contemporary of Babbage's found that a random selection of

40 volumes of tables contained 3,700 known errors. Given how important they were to the burgeoning industrial revolution, this bothered Babbage greatly.

In 1821 Babbage and his friend John Herschel got together for a fun evening rooting out mistakes. They found a lot. Babbage later wrote: 'at one point these discordances were so numerous that I exclaimed "I wish to God these calculations had been executed by steam".' At this moment Babbage hit on the idea of building an automatic computing machine.

He quickly came up with a design for a machine he called a difference engine, which could automatically perform the same calculations as a human computer, but without bodging them up.

The difference engine was so called because of the mathematical principle it was based on, the method of 'finite differences'. To cut a long equation short, this is a technique for solving mathematical expressions with two or more unknowns, such as $x = y^2 + yz - 1$. Usefully, the technique can be performed by doing nothing more than repeated additions.

In 1837, while building a prototype difference engine, Babbage started to think about a more versatile calculating machine. The difference engine could only add, but Babbage realised it should be possible to build a general-purpose machine capable of addition, subtraction, multiplication and division, which an operator could program in any sequence. He called it the Analytical Engine.

Ask me anything

The Analytical Engine is often described as the world's first computer. That is no exaggeration: it had many of the core features of modern computers, including a central processing unit and a memory. More importantly, it was capable of computing any mathematical function that is theoretically computable. In the parlance of computer science, it was 'Turing complete'.

Or it would have been if Babbage had actually built it. He worked on the design for the rest of his life, but when he died, in 1871, only part of it had been built. Perhaps it was no wonder: a full-scale Analytical Engine would have been the size of a locomotive.

It took the work of another visionary mathematician, Alan Turing, to turn such ideas into reality. In 1936, as a 24-year-old postgraduate student, Turing wrote the paper that laid the groundwork for modern computing.

He didn't intend to and was not interested in building an actual machine. Instead, he had set out to resolve a thorny but arcane mathematical challenge called the 'decision problem', posed by David Hilbert in 1928.

Hilbert wanted to know if all mathematical statements (such as $2 + 2 = 4$) could be solved, or if some were 'undecidable'. For a statement like $2 + 2 = 4$ that is trivial, but more complex ones are trickier.

If mathematics was decidable, a machine could be built to give a definite yes/no answer to any mathematical statement. All the big questions in mathematics could be resolved.

To answer the question, Turing first needed to conceptualise what sort of machine might be capable of performing such a feat. In one of the most influential thought experiments ever performed, he imagined a machine capable of reading symbols printed on an infinite strip of paper. Having read the symbol, the machine would then decide what to do next according to a set of pre-programmed rules: erase the symbol and/or write a new one; move the tape one space left or right; or stop. Depending on the rules, such a 'Turing machine' would be capable of solving mathematical problems. However, because each individual machine had fixed internal rules, it could not be used to test Hilbert's general question.

Universal machine

Then came Turing's aha moment: he realised it would be possible to define the internal rules on the tape itself. Such a device could be programmed to perform the actions of any conceivable Turing machine. It was a 'universal Turing machine', capable of performing any sequence of mathematical and logical operations. In other words, a computer.

From there it was fairly easy to work out what was and was not 'computable', and hence solve Hilbert's problem. Turing showed that there were problems that even a universal Turing machine could not solve.

It wasn't all bad news. Within five years Turing's theoretical device existed for real; the first Turing-complete computer, the Z3, was built in Berlin in 1941. The German government somehow failed to spot the military potential of the computer but the British did not make the same mistake. Turing went on to work on the Colossus computers at Bletchley Park, which played a critical role in cracking the Nazi codes.

Computers have moved on a bit since these room-filling behemoths. But under the bonnet they are merely physical actualisations of the concepts set out by Babbage and Turing.

COMPUTER SAYS NO

One of the mathematical problems that Alan Turing showed could never be solved by a computer is the self-referential halting problem, which asks 'Will this program stop?' No computer can say in advance without actually running the program. And even if it computes for a trillion years without reaching a conclusion, it still cannot say for sure. With that single result Turing proved that no procedure exists for determining whether any given mathematical statement is true or false. And so the prospect of solving all of mathematics disappeared in a puff of logic.

WHO HAD THE FIRST X-RAY VISION?

For a sober and publicity-shy man like Wilhelm Röntgen, the winter of 1895 must have been a bewildering time. It began with him wondering if he had gone mad, and ended with him as the most famous scientist in the world.

For 11 years Röntgen had been working at a provincial university in Würzburg, Germany. He was known to his colleagues as a diligent physicist, but not one marked for greatness. He certainly did not seek out the limelight. That he was unwillingly thrust into it is down to a mysterious light of a different kind that he saw glowing in his laboratory on a November evening in 1895.

Strange glow

At the time Röntgen was experimenting with a device known as a Crookes tube, an early version of the cathode ray tubes that projected images on to old-school television screens.

A Crookes tube is a partially evacuated glass chamber with an electrode at either end. Apply a high enough voltage to the electrodes and the glass opposite the negative electrode fluoresces, an effect we now know to be down to energetic electrons accelerated by the electric field – a cathode ray – hitting the atoms of the glass.

But the light Röntgen saw was nowhere near his tube. It was on the other side of his laboratory, several metres away, on a small piece of cardboard painted with a fluorescent material. That was too far for cathode rays to travel.

Röntgen started to experiment. He put the vacuum tube behind a piece of black cardboard to stop visible light, but the screen still glowed. For the next few weeks he hardly left his lab as he tried to identify the source of the glow. 'I told no one about my work,' he later wrote. 'I only informed my wife that I was doing something which, when people heard about it, would make them say Röntgen had gone mad.'

By Christmas he was sure of his own sanity. Some previously unknown type of radiation was being generated within the tube. What exactly it was, was unclear – so Röntgen called them X-rays. What was clear was that they could pass through not just cardboard but also wood and human flesh. But not bones: on one occasion Röntgen put his hand between the tube and the screen and caught a glimpse of his own skeleton.

Röntgen published his discovery in the journal of his local scientific society on 28 December 1895, to no fanfare whatsoever. But he knew he was on to something, and on New Year's Day 1896 posted reprints of his article to physicists across Europe. Twelve of them contained an insert that perhaps exposed a hidden talent for self-publicity: an X-ray photograph of his wife Anna Bertha's hand, on which her bones and wedding rings were clearly visible.

From here on things moved fast. One of the recipients of the world's first X-ray photograph was Franz Exner, a student friend of Röntgen's and now a professor of physics at the University of Vienna. A clubbable man, Exner's circle included the editor of Vienna's bestselling newspaper, *Die Presse*.

Front page news

'Eine sensationelle Entdeckung' (a sensational discovery) was the headline on its front page on 5 January 1896. It was possibly a sluggish post-Christmas news day, but they knew a scoop when they saw one – even if in their enthusiasm they misspelled Röntgen's name.

'In the learned circles of Vienna notice of a discovery that Professor Routgen is said to have made in Würzburg is presently creating a great sensation,' the article began. 'If said discovery proves to be true ... we are faced with an epoch-making result of exact science that will bring remarkable consequences in both physical and medical fields.' Newspapers around the world followed up *Die Presse*'s scoop, and X-rays became the first science-related international media sensation.

X-rays for sale

The original article correctly identified what was to become the X-ray's most important application: the capability to look into the human body. The world didn't have to wait long for X-rays to fulfil their promise. That too is largely thanks to Röntgen's modesty: he refused to cash in on his discovery, saying it 'belongs to the world at large, and should not be reserved to single enterprises through patents, licenses and the like'. Just 20 days after the initial report a Berlin firm was offering 'Röntgen tubes' for sale to physicians.

It took a while longer to work out what X-rays were. Only in 1910 was it found that they could be polarised; they were found to refract in 1912. Both pointed to X-rays being another form of electromagnetic radiation, just like light – but an extremely energetic form with a very short wavelength that allowed it to pass unhindered through many materials.

By then Röntgen was the recipient of the inaugural Nobel Prize in Physics, awarded in 1901, though he characteristically shunned the awards ceremony and donated the prize money to his university.

Röntgen died of cancer in 1923, though it is unlikely he was a martyr to his science as he was characteristically careful not to expose himself to the rays for too long. On his death his personal scientific papers were destroyed in accordance with the wishes expressed in his will.

Today few of us pass through life without coming into contact with X-rays, whether in a hospital, dentist's chair or airport security scanner. Special telescopes collect the rays and with them produce pictures of the most violent processes in the universe, such as the collisions of galaxies and the depredations of black holes. With them, we see further.

THE RELUCTANT X-MAN

Wilhelm Röntgen was a reluctant scientific celebrity but his local learned society, the Physical-Medical Society of Würzburg, was determined to promote him and his rays. A month after the discovery it held a jubilant celebration at which the president proposed to wild applause that X-rays should be renamed Röntgen Rays. With typical modesty, Röntgen stuck to his original name, as did the English-speaking world. But German and most other European languages took up the suggestion. Appropriately, Google now translates 'Herr Röntgen' as 'Mr X'.

WHAT'S LUCK GOT TO DO WITH IT?

NECESSITY, IT IS OFTEN SAID, IS THE MOTHER OF INVENTION. But sometimes it works the other way round.

In 1968, Spencer Silver of US chemical company 3M was attempting to develop a super-strong glue. He failed epically, and created a really weak one instead. But Silver noticed that it had interesting properties nonetheless. It was just strong enough to stick things together, but also weak enough to allow them to be separated without leaving a residue. It also kept its stickiness.

Silver called it his 'solution without a problem' but failed to identify the problem. He touted it around 3M to see if anyone else could find one. In 1974 a colleague called Arthur Fry attended one of Silver's seminars. Fry sang in a church choir and was annoyed that his paper bookmarks constantly fell out of his hymn book. He realised that he could use Silver's glue to temporarily secure them without damaging the pages, and from there came the idea of re-usable sticky notes.

A sticky situation

3M launched the 'Press 'n Peel' in 1977, initially to widespread indifference. But it rebranded them as 'Post-it notes' in 1980 and the product took off. The distinct colour was also an accident. Fry's lab had tons of yellow scrap paper to experiment with.

Post-it notes are sometimes cited as the quintessential example of an invention that snatched success from the jaws of failure. But they are far from the only one. Also in the serendipitous adhesive department is superglue, accidentally discovered in 1942 by chemists at Eastman-Kodak's chemical division in Rochester, New York. They were looking for clear plastics that could be made into gunsights. One day they tried a class of chemicals called cyanoacrylates, which spontaneously polymerise in the presence of water. They began testing but quickly gave up when they found that the chemicals stuck fast to everything they touched.

The team abandoned cyanoacrylates but in 1951 rediscovered them and realised their potential as a glue. At that time glues generally needed some sort of after treatment to get them to bond – pressure, heat or time – but cyanoacrylate just needed to come into contact with humidity in the air (or on your fingers). Eastman started selling it in 1958, though missed a trick by saddling it with the unsticky name Eastman #910. It sold the brand to Loctite, which did slightly better with Loctite Quick Set 404. 'Superglue' was a later coinage.

Another chemical that went from failure to fame was tetrafluoroethylene. On a Saturday morning in April 1938 a chemist called Roy Plunkett was working in his lab at DuPont in New Jersey, trying and failing to invent a new refrigerant. His starting point was an unusual gas called tetrafluoroethylene, composed of carbon and fluorine. He had a tank of the gas but couldn't get any of it out. He checked that the valve was clear. Still no gas. So he sawed the tank in half. What he found was a waxy white solid. He realised that the gas molecules had reacted with one another to create a polymer, catalysed by the steel coating of the tank.

Into the frying pan

Polytetrafluoroethylene (PTFE) turned out to have some useful properties. For one thing, the carbon–fluorine bond is incredibly strong, making the polymer extremely unreactive, which led to its first application. Scientists working on the Manhattan Project needed a way to contain fluorine, which they were using to enrich uranium. Fluorine is violently reactive but PTFE turned out to be so inert that even fluorine could not pick a fight with it, and so the pipes and valves were coated with the polymer.

Chemists soon discovered that PTFE – by then registered under the trademark Teflon – also repelled water and oil, making it a brilliant non-stick coating for frying pans. It also protected the space suits of astronauts and is nowadays used to coat heart valves because the immune system doesn't reject it.

Wartime serendipity also led to another piece of kitchen wizardry, the microwave oven. In 1945 Percy Spencer, an engineer at the US defence company Raytheon, was working on radar – the second most important military hardware project after the

Manhattan Project – when he noticed that a chocolate bar in his pocket had melted inside its wrapper.

Spencer guessed that the culprit was the electromagnetism being generated by the radar's core component, the microwave-generating cavity magnetron. So he rigged up an experimental oven and started cooking. First of all he popped some popcorn, then boiled an egg, which exploded in his colleague's face.

It turned out that the microwaves generated by the radar's cavity magnetron were just the right wavelength to set water molecules vibrating crazily and so heat them up. Two years later, Raytheon started selling microwave ovens, powered by the same cavity magnetrons used in its radar transmitters. Called the Radarange, they were much more powerful than modern microwaves, capable of cooking a baked potato in 2 minutes. 'It took many years to realise that you didn't need a radar-quality magnetron to heat food,' Spencer's grandson Rod later explained.

The age of plastic

Perhaps the most influential stroke of fortune befell Leo Baekeland, a Belgian-born chemist living in New York. In 1907 the world was gripped by a shortage of shellac, a resin secreted by insects and used as a wood preservative. He tried to create an artificial resin by combining phenol and formaldehyde but ended up with a doughy brown lump. Ever the optimist, Baekeland found that it could be moulded into shape and then set with heat to create a durable material. He had invented the world's first thermosetting plastic. Modestly he called it Bakelite, and went on to make a mint out of it. Necessity may not be the mother of invention, but fortune favours the prepared mind.

ACCIDENTS WILL HAPPEN

Post-it notes, superglue, microwaves and Teflon are the most famous examples of discoveries that unexpectedly fall out of a project aimed at trying to discover something else. But there are many, many more. When Jacob Goldenberg, an innovation researcher at the Arison School of Business in Israel, analysed the origin of 200 important inventions, he found that about half of the time, the thing was discovered before its application. More often than not, invention is the mother of necessity.

HOW DID WE BECOME THE DESTROYER OF WORLDS?

LEO SZILARD WAS WAITING TO CROSS THE ROAD NEAR RUSSELL SQUARE IN LONDON WHEN THE IDEA CAME TO HIM. It was 12 September 1933. A little under 12 years later, the US dropped an atom bomb on Hiroshima, killing an estimated 135,000 people.

The path from Szilard's idea to its deadly realisation is one of the most remarkable chapters in the history of science and technology. It features an extraordinary cast of characters, many of them refugees from Fascism who were morally opposed to the bomb but driven by the dreadful prospect of Nazi Germany getting there first.

Szilard himself was a Hungarian-born Jew who had fled Germany for the UK two months after Adolf Hitler became chancellor. He arrived in a country that was then at the forefront of nuclear physics. James Chadwick had just discovered the neutron and Cambridge physicists soon 'split the atom'. They broke a lithium nucleus in two by bombarding it with protons, verifying Einstein's insight that mass and energy were one and the same, as expressed by the equation $E = mc2$.

Szilard's eureka moment was based on this groundbreaking experiment. He reasoned that if you could find an atom that was split by neutrons and in the process emitted two or more neutrons, then a mass of this element would emit vast amounts of energy in a self-sustaining chain reaction.

Szilard pursued the idea with little success. It wasn't until 1938 that the breakthrough came – ironically in the Nazi capital Berlin, where German physicists Otto Hahn and Fritz Strassmann bombarded uranium atoms with neutrons. When they analysed the debris they were stunned to find traces of the much lighter element barium.

Chain reaction

As luck would have it, Hahn and Strassmann were opponents of the regime. Hahn wrote to the Austrian chemist Lise Meitner, who had worked with him in Berlin until she fled to Sweden after the Nazis occupied Vienna in 1938. Meitner wrote back explaining that the uranium nucleus was splitting into two roughly equal parts. She called the process 'fission'.

The next piece of the puzzle came when Italian physicist Enrico Fermi, who had fled Fascism and was working at Columbia University in New York, discovered that uranium fission released the secondary neutrons that were needed to make the chain reaction happen. Szilard soon joined Fermi in New York.

Together they calculated that a kilogram of uranium would generate about as much energy as 20,000 tonnes of TNT. Szilard already saw the prospect of nuclear war. 'There was very little doubt in my mind that the world was headed for grief,' he later recalled.

Others did have doubts, however. In 1939 the Danish physicist Niels Bohr – who was actively helping German scientists escape via Copenhagen – poured cold water on the idea. He pointed out that uranium-238, the isotope which makes up 99.3 per cent of natural uranium, would not emit secondary neutrons. Only a very rare isotope of uranium, uranium-235, would split in this way.

However, Szilard remained convinced that the chain reaction was possible, and feared that the Nazis knew it too. He consulted fellow Hungarian émigrés Eugene Wigner and Edward Teller. They agreed that Einstein would be the best person to alert President Roosevelt to the danger. Einstein's famous letter was sent soon after the outbreak of war in Europe, but had little impact.

Things changed dramatically in 1940, when news filtered through that two German physicists working in the UK had proved Bohr wrong. Rudolf Peierls and Otto Frisch had worked out how to produce uranium-235 in large quantities, how it could be used to produce a bomb, and what the appalling consequences of dropping it would be. Peierls and Frisch – who Bohr had helped escape – were also horrified at the prospect of a Nazi bomb, and in March they wrote to the British government urging prompt action. Their 'Memorandum on the Properties of a Radioactive "Super-Bomb"' was more successful than Einstein's letter to Roosevelt. It led to the initiation of the British bomb project, codenamed Tube Alloys.

The letter also galvanised the US into action. In April 1940 the government appointed the veteran physicist Arthur Compton to head a nuclear weapons programme, which eventually became the Manhattan Project. One of his first moves was to bring together various chain reaction research groups under one roof in Chicago. That summer the team began a series of experiments to make the chain reaction happen.

The bombing of Pearl Harbor in December 1941 added further impetus. A year later the Manhattan Project team was ready to attempt a chain reaction in a pile of uranium and graphite they had assembled in a squash court underneath a stand of the University of Chicago's football field. On Wednesday, 2 December 1942, they did it.

A black day

Celebrations were muted. Once the reaction was confirmed, Szilard shook hands with Fermi and said: 'This will go down as a black day in the history of mankind.'

Over the next four years the US, UK and Canada poured vast resources into the Manhattan Project. Tube Alloys continued for a while but was eventually absorbed into the US project. The Nazis initiated a nuclear weapons programme but made little progress.

On 16 July 1945 the US detonated the world's first nuclear bomb in the New Mexico desert. The test was final, terrible proof that nuclear energy could be weaponised, and prompted Robert Oppenheimer to recall a passage from the Hindu scripture, Bhagavad Gita: 'I am become death, the shatterer of worlds.'

The attacks on Japan started a worldwide arms race. Following

1945, the US developed massively destructive hydrogen bombs, which exploited nuclear fusion rather than fission. The Soviets developed and tested their own bomb in 1949. The world's nuclear arsenal now stands at 27,000 bombs.

BOMBSHELL MOMENT

The Manhattan Project was spurred on by the fear that the Nazis would win the race to build the bomb. We now know that was never going to happen. After Germany surrendered in 1945, 10 of its leading nuclear scientists were interned in a country house near Cambridge. The rooms were bugged and the transcripts leave no doubt that the Germans were nowhere near an atom bomb and did not believe one was possible.

HOW DID WE BEAT THE BUGS (FOR A WHILE)?

'WHEN I WOKE UP JUST AFTER DAWN ON SEPTEMBER 28, 1928, I CERTAINLY DIDN'T PLAN TO REVOLUTIONISE ALL MEDICINE ... But I suppose that was exactly what I did.' This is how Alexander Fleming described his discovery of penicillin, one of the greatest breakthroughs in biomedical science.

The official version of events is well known. Fleming was a microbiologist at St Mary's Hospital in London investigating a group of disease-causing bacteria called *staphylococci*, or staph for short. Returning early from a holiday, he noticed that one of his culture dishes had been contaminated by a mould that was stopping the staph from growing. He later speculated that spores of the mould had blown in through an open window.

Mould broth

Fleming cultured the mould in a broth and found that an extract killed a number of disease-causing bacteria, especially those responsible for diphtheria – though it had no effect on many others including typhoid and cholera. At first he called the liquid mould broth filtrate but later renamed it penicillin after the mould's Latin name, *Penicillium*. He also showed that penicillin was non-toxic to animals, even at enormous doses.

Fleming published his results in a paper in 1929 in which he also suggested that penicillin might be used to treat bacterial infections. Over the next decade he embarked on a tireless crusade to realise its potential. Most important of all, he sought chemists to extract and purify it in large quantities – something Fleming himself never succeeded in doing.

Eventually a team at the University of Oxford, led by the Australian Howard Florey, cracked the problem, and by 1944 penicillin was being mass produced. Its use on casualties of the D-Day landings cemented its place in medical history.

Penicillin was undoubtedly the medical breakthrough of the twentieth century. Before, 80 per cent of people with blood poisoning caused by staph infections died; after, almost none did. It is impossible to say how many lives it saved, but tens of millions seems a conservative estimate. And the antibiotic age which it ushered in saved hundreds of millions more. But the real story of its discovery is not quite as triumphant as it seems.

Boring, boring

For one thing, the antibacterial effects of the *Penicillium* mould were well known before Fleming. And though he went further down the road to exploiting this property than anyone before, he almost dropped the ball completely.

Fleming's 1929 paper was widely ignored. Fleming also gave a number of presentations, but they created no stir because he was a boring speaker. He continued to work on penicillin but it was not his priority and the research went badly. One experiment showed that once injected into mice, penicillin disappeared from the bloodstream after 30 minutes, whereas it needed 4 hours to kill bacteria in a dish. This seems to have convinced Fleming that it probably wouldn't work.

If discovery is 1 per cent inspiration and 99 per cent perspiration, then Florey's group deserves most of the credit. His was one of many teams that had independently picked up on Fleming's work and were attempting to turn penicillin into a therapy. But aside from supplying them with cultures of the mould to work with, Fleming paid them no attention. He only became interested in 1942 once they had succeeded.

Oh what a lovely war

Florey had struggles of his own. He started out in 1938, but kept on running short of money. His team eventually succeeded in separating out penicillin and demonstrated that it could be used to treat infections. But in spite of using every available container, including dustbins and bedpans, they could never make enough of the stuff to convince governments or private firms to start up large-scale production.

In the middle of their travails, they had a stroke of luck: the

Second World War broke out. The US and UK governments poured money into the project and eventually mass production of penicillin was arranged in the US.

So how did Fleming end up taking so much of the credit? The answer is that he had better PR. The bandwagon was set rolling by the newspaper mogul Lord Beaverbrook, a patron of St Mary's Hospital, who arranged for glowing coverage of Fleming's discovery. It was given further impetus by Almroth Wright, Fleming's boss at St Mary's, who wrote to *The Times* claiming primary credit for Fleming. Wright was motivated by the PR value: like all British teaching hospitals at the time, St Mary's was dependent on charitable donations.

Incidentally, Florey later received a begging letter from St Mary's which began 'You may have heard of the discovery of penicillin by Dr Alexander Fleming'. Florey had it framed and hung on his wall. The UK Ministry of Information also perpetuated the Fleming myth for propaganda purposes.

In 1945, Fleming, Florey and Florey's right-hand man Ernst Chain were jointly awarded the Nobel Prize, but Fleming soaked up most of the limelight. Between being awarded the prize and his death in 1955, he received 140 other major awards. After his death his notoriously untidy laboratory was turned into a museum.

To be fair to Fleming, when he spoke to reporters he invariably directed them to Oxford to get Florey's side of the story. But Florey refused to speak to the press and forbade his team from doing so. Reporters were also more drawn to the story of Fleming's serendipitous discovery than Florey's methodical graft – bedpans and dustbins notwithstanding. And so the penicillin myth took on a life of its own.

THE SWINGING FIFTIES

The sexual revolution is often attributed to the invention of the contraceptive pill, which became available in the US in 1960. But another drug probably got the party started much earlier: penicillin. In 1939 syphilis killed 20,000 people in the US and gonorrhoea was also rife. By the mid 1950s penicillin had almost wiped these diseases out – an event which neatly coincided with a sudden shift in public attitudes towards casual sex.

DID THE GEEKS REALLY INHERIT THE EARTH?

TRY TO IMAGINE LIFE BEFORE THE INTERNET. No smartphones. No social media. No Google. No Netflix, Spotify, Amazon, Uber or Airbnb. Not even email. If you wanted to read the news, you bought a newspaper. If you wanted to listen to music, you bought a CD. If you wanted to communicate with somebody who wasn't in earshot, you phoned them up.

How did we survive all those years ago, back in the Dark Ages, back in 1990?

The internet is so pervasive today that it is easy to forget how new it still is. Twenty years ago only about half of all Americans had even heard of it; even then, it was essentially an arcane and technical computer science project whose pioneers probably had no idea what it would end up being used for, or how transformative it would be.

A better way to connect

If there is a year zero for the internet, it is probably 1961. Back then, communications systems were direct channels from one location to another. Telephone calls relied on a physical line between two telephones. Radio communications were broadcast from one point to another. But direct links are hugely inefficient. That computers could be connected in a better way was the key insight that made the internet possible. It was the result of work by Leonard Kleinrock, an engineer at the Massachusetts Institute of Technology, who in 1961 had begun thinking about how data could best flow through large networks of computers.

Instead of the entire message travelling directly from one point to another, Kleinrock's insight was to chop the message into pieces, or packets, and allow each to find its own way through the network. The destination computer would then reassemble the message once all the packets had arrived.

'Packet switching' proved to be much more efficient and flexible than using dedicated lines. If a link between two computers went down, the packets could just find another route. But it required a major rethink of the way communication networks operated. It needed devices on the network to read each packet and route them towards their destination. And it needed a special piece of code attached to each packet that told the router what the message was and how to reassemble it. This code later evolved into a set of rules called the internet protocol, including a dedicated address for each computer on the network – the IP address.

In 1966 the work came to the attention of a military R&D organisation called the Advanced Research Projects Agency (ARPA), which asked Kleinrock to create a large-scale computer network to connect up its researchers. By then, Kleinrock had moved to the University of California, Los Angeles. He set up the first node in his laboratory there and the second at the Stanford

Research Institute near San Francisco. More nodes were added and the growing network became known as ARPANET.

Crash and burn

The first transmission didn't go well. Kleinrock's computer crashed while sending the word 'login' so the first message was simply 'lo'. The complete word was successfully sent an hour or so later. That was 29 October 1969.

By 1973 ARPANET reached from Hawaii, across the US to London. As it grew it became clear that its control software wasn't up to the job. Two computer scientists, Vint Cerf and Robert Khan, produced a better version. They updated the internet protocol to create a set of rules called TCP/IP (Transmission Control Protocol/Internet Protocol), which detailed everything from how computers should identify one another to the detection of transmission errors.

In 1975 they successfully tested TCP/IP over a link between Stanford University and University College London. It was a huge moment in internet history, but there was trouble brewing. Computer networks had sprung up around the world but most used their own communication rules. These networks could not talk to each other and the internet was in danger of becoming a Tower of Babel, with everyone speaking different languages.

That changed, slowly. In 1982 the US Department of Defense adopted TCP/IP across all its networks and the next year ARPANET followed. Meanwhile, the telephone company AT&T

began developing a version of TCP/IP written in the UNIX computer language. Crucially, it placed this code in the public domain for anybody to use.

This act of enlightened generosity had a major impact on the spread of the internet, since any computer running UNIX could get online. This was 1989; it is no coincidence that the internet's extraordinary explosive growth followed soon after.

Another significant development took place the same year. At the time, the largest internet node in Europe was at CERN, the particle physics laboratory near Geneva. There, a young computer scientist called Tim Berners-Lee was frustrated by the internet's lack of a system for viewing, sharing and linking to documents. To solve these problems Berners-Lee created a software system called WorldWideWeb, which included the first ever Web browser. It also included a feature for creating hyperlinks called hypertext transfer protocol or HTTP. He used it to build the first website and placed it online at info.cern.ch.

Information revolution

WorldWideWeb was the software that the internet – the hardware – needed to break out of the lab. It spread like wildfire, piggybacking on the spread of the infrastructure. In 1993 the network carried just 1 per cent of the world's information flow. Today it is close to 100 per cent, a technological revolution of unprecedented scale and speed. If you do remember the days before the internet, consider yourself fortunate: you witnessed history in the making.

THE INTERNET, 1968 STYLE

Another landmark event in the evolution of the internet has become known as The Mother of all Demos. On 8 December 1968 a team of engineers from the Stanford Research Institute in California gathered at a computer technology conference in San Francisco to demonstrate the future of computing. Among other innovations they unveiled video conferencing, collaborative editing, hypertext and the computer mouse. This visionary presentation glimpsed a way of life that would become normal some 30 years later, but only after the explosive growth of an entirely new system of communication called the internet.

HOW DID WE CONQUER SPACE?

On 8 September 1944 the world woke up to a terrifying new weapon. First Paris and then London were hit by giant flying bombs raining down from the sky. For Nazi Germany the V-2 ballistic missile was a last throw of the dice. Hitler believed it would turn the tide of the war he was losing. He was wrong, but it did change the course of history.

The V-2 was not the first rocket weapon: gunpowder-fuelled missiles were invented during the Napoleonic Wars. But it was the first rocket with the oomph to travel high into the atmosphere and get close to the edge of space.

For that we have to thank an American engineer, Robert Goddard, born in 1882. He taught himself aerodynamics during a bout of childhood illness and later became convinced that space flight was possible.

In 1914 Goddard filed two patents for what he realised was the only technology powerful enough to escape Earth's gravity: multi-staged, liquid-fuelled rockets. In 1919 he expanded on his ideas in a seminal work, *A Method of Reaching Extreme Altitudes*.

Rocket men

Goddard wasn't the only engineer with designs on space. In 1922 a German called Hermann Oberth submitted a PhD thesis on rocket science to the University of Heidelberg. It was rejected. But the next year he self-published a book called *By Rocket into Planetary Space*, which inspired a group of like-minded Germans to form the Society for Space Travel.

Meanwhile, the Soviet Union set up the Society for Studies of Interplanetary Travel, an official body spun out of the military academy in Moscow. In October 1924 the society held a public debate about the feasibility of launching a rocket to the Moon. The race was on to produce the first liquid-fuelled rocket, and ultimately to escape the bonds of Earth's gravity.

First blood went to the Americans. On 16 March 1926 Goddard oversaw the first liquid-fuelled rocket launch, in Auburn, Massachusetts. It wasn't exactly shooting for the Moon: the rocket flew for 2.5 seconds, reached an altitude of 12 metres and crash-landed in a cabbage field. Realising he needed some means of steering his craft, Goddard added moveable vanes and gyroscopic control.

At this point Goddard was well ahead of the game. But his rivals were catching up. In 1929 Oberth successfully demonstrated a rocket engine in a static test. His team included an 18-year-old student called Wernher von Braun, who quickly overtook Oberth as the de facto leader of the German efforts.

In 1933 the Soviet Union carried out its own test launch under the guidance of another future colossus of rocket science, Sergei Korolev. He would become the guiding light of the USSR's space programme and was still in charge when Yuri Gagarin became the first man to orbit the Earth in 1961.

As the world mobilised in anticipation of war, governments and their armed forces began to take increasing interest in rockets. Imagine being able to launch explosives towards another nation at the press of a button! Loftier ideals about space flight were put on the back burner.

Bombs away

Germany quickly took the lead. In 1933 work began on the prototype that would lead to the V-2; the first successful test launches happened in 1934. But after early progress, von Braun and his team hit a series of snags, not least that Hitler was unenthusiastic. But as the war swung the way of the Allies, the programme was stepped up.

By some measures the V-2 was a great success. It was the world's first ballistic missile and, more pertinently, the first human-made object to reach space, in a test flight on 20 June 1944. Engineers were now in little doubt that large liquid-propellant rocket engines were capable of taking humans into space.

The defeat of the Nazis put Germany out of the space race, but its rocket scientists continued to work on the problem after being headhunted by their former enemies in the US and USSR. In the first instance both sides wanted unused V-2s and the technology to build more. Later they were keen to build intercontinental ballistic missiles to launch nuclear warheads. Eventually they started a race to the Moon. All of it was based on German rocketry.

The competition was stiff and the progress rapid. In 1946 a camera atop a V-2 launched from White Sands Missile Range in New Mexico captured an image of the curvature of the Earth and the void beyond, the first photograph from space. It was around this time that the expression 'it's not rocket science' entered the vernacular.

Both sides also made improvements on the V-2, building bigger and better rockets. It was just a matter of time before one side or the other was able to launch a body into orbit. That day came on 4 October 1957.

It's difficult now to comprehend the consternation that the Soviet Union's launch of the world's first artificial satellite created. The west watched with trepidation as Sputnik 1 bleeped a feeble radio signal back to Earth. The small metal sphere remained aloft for 10 weeks before burning up on re-entry. Considering it was only 58 centimetres in diameter with nothing more sinister inside than a radio transmitter, the concern seems disproportionate. Nonetheless, the US had been beaten into orbit. Four years later the Soviets seemingly doubled their advantage when Korolev launched Gagarin into orbit.

Walking on the Moon

But the US had the last laugh. In August 1969 NASA used a Saturn V rocket to put humans on the Moon. What had started as a last ditch bid to win the war for Germany ended up delivering a propaganda victory for the United States – and helped fuel a space research programme that has taught us more than any other about where we came from, and where we are going.

LIFE IMITATES SCI-FI

Space has always exerted a powerful tug on the imagination, but realistic dreams of actually going there only began in the 1860s when Jules Verne published the novels *From Earth to the Moon* and *Around the Moon*. Both Robert Goddard, the father of American rocket science, and his German counterpart Hermann Oberth found boyhood inspiration in these works of science fiction, and, despite widespread scepticism, helped to make them come true a century after they were published.

FURTHER READING

PART ONE: THE UNIVERSE

Matter, space and time

A Brief History of Time: From the Big Bang to Black Holes by Stephen Hawking (Bantam Dell, 1988)

Stars and galaxies

Galaxies: A Very Short Introduction by John Gribbin (Oxford University Press, 2008)

Chemical elements

The Elements: A Visual Exploration of Every Known Atom in the Universe by Nick Mann and Theodore Gray (Black Dog & Leventhal, 2011)

Meteorites

Atlas of Meteorites by Monica M. Grady, Giovanni Pratesi and Vanni Moggi Cecchi (Cambridge University Press, 2013)

Dark matter and dark energy

The 4% Universe: Dark Matter, Dark Energy, and the Race to Discover the Rest of Reality by Richard Panek (Oneworld, 2012)

Black holes

Black Holes: The Reith Lectures by Stephen Hawking (Bantam, 2016)

PART TWO: OUR PLANET

The solar system

Wonders of the Solar System by Brian Cox and Andrew Cohen (Collins, 2010)

The Moon

The Moon, a Biography by David Whitehouse (Orion, 2002)

Continents and oceans

Ocean Worlds: The Story of Seas on Earth and Other Planets by Jan Zalasiewicz and Mark Williams (Oxford University Press, 2014)

Weather

The Cloudspotter's Guide by Gavin Pretor-Pinney (Sceptre, 2006)

Soil

Earth Matters: How Soil Underlies Civilization by Richard Bardgett (Oxford University Press, 2016)

Air

Out of Thin Air: Dinosaurs, Birds, and Earth's Ancient Atmosphere by Peter Ward (National Academies Press, 2006)

Oil

The Prize: The Epic Quest for Oil, Money and Power by Daniel Yergin (Simon & Schuster, 1991)

PART THREE: LIFE

Life

Creation: The Origin of Life / The Future of Life by Adam Rutherford (Penguin, 2014)

Complex cells

The Vital Question: Why is Life The Way It Is? by Nick Lane (Profile Books, 2015)

Sex

Power, Sex, Suicide: Mitochondria and the Meaning of Life by Nick Lane (Oxford University Press, 2005)

Insects

Planet of the Bugs: Evolution and the Rise of Insects by Scott Shaw (University of Chicago Press, 2014)

Dinosaurs

Dinosaurs by Michael Benton and Steve Brusatte (Quercus, 2008)

Eyes

Climbing Mount Improbable by Richard Dawkins (W. W. Norton, 1996)

Sleep

Sleep: A Very Short Introduction by Steven W. Lockley and Russell G. Foster (Oxford University Press, 2012)

Humans

The Strange Case of the Rickety Cossack and Other Cautionary Tales from Human Evolution by Ian Tattersall (Palgrave Macmillan, 2015)

Language

The Evolution of Language by W. Tecumseh Fitch (Cambridge University Press, 2010)

Friendship

How Many Friends does One Person Need?: Dunbar's Number and Other Evolutionary Quirks by Robin Dunbar (Faber & Faber, 2010)

Belly button fluff

Elephants on Acid: And Other Bizarre Experiments by Alex Boese (Mariner Books, 2007)

PART FOUR: CIVILISATION

Cities

Mesopotamia: The Invention of the City by Gwendolyn Leick (Penguin, 2002)

Money

Money Changes Everything: How Finance Made Civilization Possible by William N. Goetzmann (Princeton University Press, 2016)

Funerals

The Palaeolithic Origins of Human Burial by Paul Pettitt (Routledge, 2010)

Cooking

Catching Fire: How Cooking Made us Human by Richard Wrangham (Profile Books, 2010)

Domesticated animals

The Covenant of the Wild: Why Animals Chose Domestication by Stephen Budiansky (Orion, 1994)

Organised religion

Big Gods: How Religion Transformed Cooperation and Conflict by Ara Norenzayan (Princeton University Press, 2015)

Alcohol

Uncorking the Past: The Quest for Wine, Beer, and Other Alcoholic Beverages by Patrick E. McGovern (University of California Press, 2009)

Possessions

Paraphernalia: The Curious Lives of Magical Things by Steven Connor (Profile Books, 2011)

Clothes

The Wild Life of our Bodies: Predators, Parasites, and Partners that Shape Who We Are Today by Rob Dunn (HarperCollins, 2011)

Music

The Singing Neanderthals: The Origins of Music, Language, Mind and Body by Steven Mithen (Harvard University Press, 2006)

Personal hygiene

Bum Fodder: An Absorbing History of Toilet Paper by Richard Smyth (Souvenir Press, 2012)

PART FIVE: KNOWLEDGE

Writing

Lost Languages: The Enigma of the World's Undeciphered Scripts by Andrew Robinson (McGraw-Hill, 2002)

Zero

Nothing: From Absolute Zero to Cosmic Oblivion – Amazing Insights into Nothingness by *New Scientist* (Profile Books, 2013)

Measurement

The Measure of All Things: The Seven-Year Odyssey and Hidden Error that Transformed the World by Ken Alder (Little, Brown, 2002)

Timekeeping

The Mastery of Time: A History of Timekeeping, from the Sundial to the Wristwatch. Discoveries, Inventions, and Advances in Master Watchmaking by Dominique Fléchon and Franco Cologni (Flammarion, 2011)

Politics

The Righteous Mind: Why Good People are Divided by Politics and Religion by Jonathan Haidt (Penguin, 2013)

Chemistry

The Disappearing Spoon: And Other True Tales of Madness, Love, and the History of the World from the Periodic Table of the Elements by Sam Kean (Little, Brown, 2010)

Quantum mechanics

Quantum Theory Cannot Hurt You: Understanding the Mind-Blowing Building Blocks of the Universe by Marcus Chown (Faber & Faber, 2014)

PART SIX: INVENTIONS

The wheel

The Wheel: Inventions and Reinventions by Richard W. Bulliet (Columbia University Press, 2016)

Radio

Marconi: The Man who Networked the World by Marc Raboy (Oxford University Press, 2016)

Flight

First Flight: The Wright Brothers and the Invention of the Airplane by T. A. Heppenheimer (John Wiley, 2003)

Qwerty keyboard

Quirky Qwerty: A Biography of the Typewriter & its Many Characters by Torbjorn Lundmark (Penguin, 2003)

Computers

Alan Turing: The Enigma by Andrew Hodges (Princeton University Press, 2014)

X-rays

Röntgen Rays: Memoirs by Wilhelm Conrad Röntgen, Sir George Gabriel Stokes and Sir Joseph John Thomson (Sagwan Press, 2015)

Accidental discoveries

Chance: The Science and Secrets of Luck, Randomness and Probability by *New Scientist* (Profile Books, 2015)

Nuclear weapons

Inside the Centre: The Life of J. Robert Oppenheimer by Ray Monk (Jonathan Cape, 2012)

Antibiotics

Alexander Fleming: The Man and the Myth by Gwyn Macfarlane (Chatto & Windus, 1984)

The internet

Tubes: Behind the Scenes at the Internet by Andrew Blum (Ecco Press, 2012)

Rocket science

Rockets into Space by Frank H. Winter (Harvard University Press, 1993)

ACKNOWLEDGEMENTS

This book would not have come about without the support of many people at New Scientist, especially Sumit Paul-Choudhury and John MacFarlane. Thanks to Catherine Brahic, Daniel Cossins, Liz Else, Dave Johnston, Will Heaven, Valerie Jamieson, Frank Swain and Jeremy Webb for their ideas and suggestions, and to everybody else at New Scientist for their general brilliance.

Thanks also to the equally brilliant team at John Murray: Nick Davies, Georgina Laycock and Kate Miles in editorial, Amanda Jones in production, Rosie Gailer in publicity, Ross Fraser in marketing, Al Oliver in art and Ben Gutcher in sales.

Meanwhile in San Francisco, thanks to Alan McLean for his critical eye, Derek Watkins for his mapping expertise and Brian X. Chen for his earwax. Many of these illustrations would not be possible without the use of D3, a JavaScript library for visualising data.

Some of the material in this book is adapted from articles published previously in *New Scientist*.

INDEX